HOLOGRAPHICOSMIC MAN

The Holographic Heart of the Golden Mean

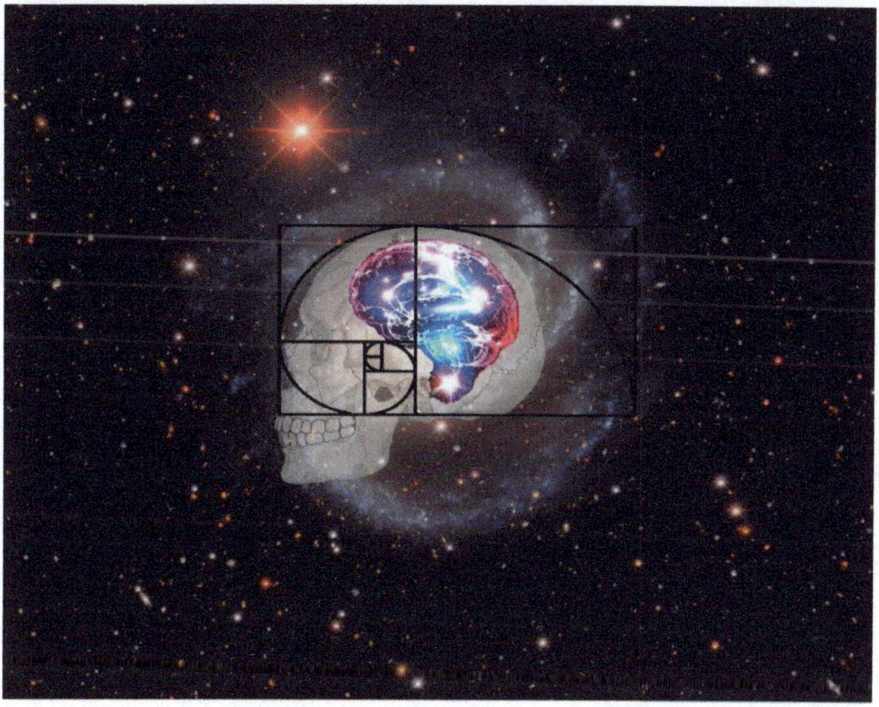

By

Matthew J. Pallamary

Mystic Ink Publishing

Mystic Ink Publishing
San Diego, CA
www.mysticinkpublishing.com

ISBN 10: 0-9986809-8-2 (sc)
ISBN 13: 978-0-9986809-8-9 (sc)

Library of Congress Control Number: 2022910843
Mystic Ink Publishing, San Diego, CA

Book Jacket and Page Design: Matthew J. Pallamary/ San Diego CA
Author's Photographs: Matthew J. Pallamary -- Gibbs Photo/Malibu CA

Acknowledgements

The author would like to thank Scott Olsen Ph.D., Professor of Comparative Religion for sharing his research and passion for the Golden Mean, to Science Fiction writer David Brin, a valued friend who "gets me", Ken Symington for his mentorship, Stanley Krippner Ph.D. for his unflagging support and friendship, and Jacques Olivier for his proofing skills. Last but not least is my infinite gratitude for my greatest inspiration of them all; M.A.

TABLE OF CONTENTS

WHAT'S THE MATTER?

The idea of creating matter from light stems from Einstein's famous $E=MC^2$ equation which states that energy and matter are interchangeable, but using light energy to prove that photons are real and long-lived, and not "virtual" is challenging. Although today's most powerful lasers can't produce matter, photons surrounding ions at the Relativistic Heavy Ion Collider can.

Nuclear scientists used the particle accelerator to create matter from collisions of light by accelerating two beams of gold ions close to the speed of light in opposite directions. At these speeds each gold ion is surrounded by particles of light which are photons generated by the ion's perpendicular magnetic and electric fields. When the ions grazed past each other without colliding, the photons interacted producing electrons (matter) and positrons (antimatter).

Accelerating heavy ions like gold generates a magnetic field similar to current flowing through a wire. At high speed the spiraling magnetic and perpendicular electric fields are of equal strength which is the definition of a photon; a quantized "particle" of light.

This transformation of light defined by waves and measured in wavelengths into matter defines matter as a substance that has inertia and occupies physical space, and it goes to the heart of the concept of matter waves, a central part of the theory of quantum mechanics and an example of wave–particle duality.

All matter exhibits wave-like behavior.

In accordance with this concept, a beam of electrons can be diffracted just like a beam of light or a water wave.

Wave–particle duality in quantum mechanics states that every particle or quantum entity can be described as either a particle or a wave. Albert Einstein stated, "It seems as though we must use sometimes the one theory and sometimes the other, while at times we may use either. We are faced with a new kind of difficulty. We have two contradictory pictures of reality; separately neither of them fully explains the phenomena of light, but together they do."

Current scientific theory holds that all particles exhibit a wave nature and vice versa, a phenomenon that has been verified for both elementary and compound particles like atoms and molecules. Although the use of the wave–particle duality has worked well in physics, its meaning has not been satisfactorily resolved.

Niels Bohr, a Danish physicist who received the Nobel Prize in Physics for his contributions to understanding atomic structure and quantum theory regarded this "duality paradox" as a fundamental metaphysical fact of nature. This paradox states that a given quantum object will sometimes appear as a wave, and sometimes as a particle in different physical settings. Bohr saw this duality as one aspect of the concept of complementarity which states that objects have certain pairs of complementary properties which cannot be observed or measured simultaneously.

An example of such a pair is position and momentum. Bohr considered one of the foundational truths of quantum mechanics to be the fact that setting up an experiment to measure one quantity of a pair, such as the position of an electron, excludes the possibility of measuring the other, yet understanding both experiments is necessary to characterize the object under study. In Bohr's view, the behavior of atomic and subatomic objects cannot be separated from the measuring instruments that create the context in which the measured objects behave. Consequently, there is no "single picture" that unifies the results obtained in these different experimental contexts and only the "totality of the phenomena" together can provide a complete informative description. Bohr regarded renunciation of the cause-effect relation, or complementarity of the space-time picture as essential to the quantum mechanical account.

Around the same time, Werner Heisenberg, a German theoretical physicist saw this duality as present for all quantic entities, but not in the same perspective as Bohr, giving rise to Heisenberg's uncertainty principle that he explained as a consequence of the process of measuring that disturbed momentum and vice versa.

The uncertainty principle put forth by Bohr and Heisenberg depends more on the wave nature of a particle. In the same way that it is nonsensical to discuss the precise location of a wave on a string, particles do not have precise positions. It is also nonsensical to discuss the wavelength of a "pulse" wave traveling down a string. Particles do not have precise momenta that correspond to the inverse of wavelength. When position is well defined, the wave is pulse-like and has a poorly defined wavelength and corresponding momentum. Conversely, when momentum, and by association wavelength is well defined, the wave looks long and sinusoidal, giving it a poorly defined position. Wave–particle duality is deeply embedded into the foundations of quantum mechanics.

The formalism view holds that statements of mathematics and logic can be considered to be statements about the consequences of the manipulation of strings. In this theory all the information about a particle is encoded in its wave function that is analogous to the amplitude of a wave at each point in space. Propagation of these waves leads to wave-like phenomena like interference and diffraction.

Measuring the location of a particle forces it into a more localized state given by the uncertainty principle. When viewed through this formalism, the measurement of the wave function will randomly lead to wave function collapse into a sharply peaked function at some location.

Wave–particle duality is an ongoing conundrum in modern physics. Most physicists accept it as the best explanation for a broad range of observed phenomena, but it is not without controversy.

According to wave only theory, quanta are separate entities that evolve and interact according to deterministic equations, except when a quantum transfers its energy to an absorbing atom, then it disappears from all of space. What is called a particle is considered a holistic unit of field that in its totality behaves as a particle. Even if it is spread out over many kilometers it has to disappear instantaneously, like a photon collapsing into a photoreceptor cell in the eye. While instantaneous collapse is hard for many physicists to accept, there is nothing logically inconsistent about it, and it does not violate the Principle of Relativity.

It has been argued that there are never exact particles or waves, but only some compromise or intermediate between them. For this reason, in 1928 Arthur Eddington coined the name "*wavicle*" to describe these objects. One consideration is that zero-dimensional mathematical points cannot be observed. Another is that the formal representation of such points is unphysical, because they cannot be normalized. Parallel arguments apply to pure wave states.

Nobel Laureate Roger Penrose, a British mathematician, mathematical physicist, and a philosopher of science stated. "Such 'position states' are idealized wavefunctions in the opposite sense from the momentum states. Whereas the momentum states are infinitely spread out, the position states are infinitely concentrated. Neither is normalizable..."

This ongoing enigma regarding the elusive definition of our experience of reality goes to the heart of the subjective objectivity "interference pattern" that defines our perspective and subsequent perception of the world as we know it.

By definition, an objective statement is based on facts and observations. On the other hand, a subjective statement relies on assumptions, beliefs, and opinions influenced by emotions and personal feelings, but how can we define anything as truly objective when the act of observation which is subjective in itself alters that very observation?

The timeless expression, "beauty is in the eye of the beholder" also holds true with the statement defined by this paradox. Substituting the word beauty to say, "reality is in the eye of the beholder", gives us the definition of radical subjectivity, the way each of us sees the world in the definitive "reality" of our own unique individual perspectives. This indicates that the "solid world" that we believe we exist in is in fact not so solid after all, but is more ephemeral and illusory than we think, something the ancients called **maya**, a Sanskrit word meaning "magic" or "illusion".

Maya is a fundamental concept in Hindu philosophy, notably in the Advaita nondualist school of Vedanta. Maya originally denoted the magic power with which a god can make human beings believe in what turns out to be illusion. It later came to mean *the* powerful force that creates the cosmic illusion that the phenomenal world is real.

For nondualists, maya is the cosmic force that presents the infinite *brahman* known as the supreme being, the finite phenomenal world defined as the infinite nested within the finite, or conversely the finite nested within the infinite; a manifestation of the implicate/explicate order characteristic of holograms.

Maya is reflected on the individual level by human ignorance of the real nature of the self which is mistaken for the empirical ego, which is identical with *brahman*. This ancient concept of maya is defined as a *powerful force* that creates the illusion we think of as "reality".

In its simplest terms force is defined as strength or energy as an attribute of physical action or movement, or put another way, force is an expression of energy, and power is the rate at which energy is transferred, or the rate at which work is done.

Einstein's $E=MC^2$ equation stating that energy and matter are interchangeable combined with the fact that all matter exhibits wave-like behavior not only validates the ancient nondualist belief in maya, but every element of this statement taken as a whole indicates that all of these multifaceted aspects are in fact manifestations of one and the same phenomenon.

EVERYTHING IS ENERGY

Nikola Tesla is quoted as saying, "If you want to find the secrets of the universe, think in terms of energy, frequency, and vibration."

Another quotation often attributed to Albert Einstein, is widely believed to come from a channeler who assigned the words to an entity named Bashar, who stated, "Everything is energy and that's all there is to it. Match the frequency of the reality you want and you cannot help but get that reality. It can be no other way. This is not philosophy. This is physics."

Einstein spoke about the relationship between matter and energy in a 1948 film called "Atomic Physics", where he stated, "It followed from the special theory of relativity that mass and energy are both but different manifestations of the same thing — a somewhat unfamiliar conception for the average mind."

Many years earlier, George Ivanovich Gurdjieff, a Russian philosopher, mystic, spiritual teacher, and composer of Armenian and Greek descent, born in 1866 stated the same concept in his own way.

"In the Absolute, or all is one, matter and force are also one. But in this connection matter and force are not taken as real principles of the world in itself, but as properties or characteristics of the phenomenal world observed by us. To begin the study of the universe it is sufficient to have an elementary idea of matter and energy, such as we get by immediate observation through our organs of sense. The 'constant' is taken as material, as matter, and 'changes' in the state of the 'constant', or of matter, are called manifestations of force or energy. All these changes can be regarded as the result of vibrations or undulatory motions which begin in center, that

is, in the Absolute, and go in all directions, crossing one another, colliding, and merging together, until they stop all together at the end of the ray of creation.

From this point of view, then, the world consists of vibrations and matter, or of matter in a state of vibration, a vibrating matter. The rate of vibration is in inverse ratio to the density of matter."

The word *energy* is derived from the Ancient Greek expression: ἐνέϱγεια *energeia* "activity, operation", which is believed to appear for the first time in the work of Aristotle. In contrast to the modern definition, energeia was a qualitative philosophical concept broad enough to include ideas like happiness and pleasure. Every emotion that we feel, we experience as a different kind of energy, and each has its own unique qualities.

In physics, energy is a property of objects, transferable among them through fundamental interactions, which can be converted in form, but not created or destroyed. Energy transformation or energy conversion is the process of changing one form of energy to another indicating movement, and movement denotes transformation. The geometric concept of a point is a coordinate that has no dimension which is static and unmoving. If you extend that point in one direction, you have a line. The more energy that is expended overcoming inertia, the further the point moves and the longer the line becomes, so a longer line is a sign of more spent energy.

The distance between these two points contains the potential, which in the study of electricity is called voltage, electrical potential difference, electric tension, or electric pressure, which in essence is the potential difference between two points.

Implicit in the difference between two points is the concept of polarity denoted by two poles labeled positive and negative. Batteries, fuel cells and solar cells all produce this direct current, known as DC. The terminals of a battery are always positive and negative and current always flows in the same direction between those two terminals.

The power that comes from a power plant is supplied in alternating current, known as AC where the direction of the current reverses or alternates 60 times per second in the U.S. or 50 times per second in Europe.

There are two characteristics of alternating current that come into play with this movement known as amplitude and frequency.

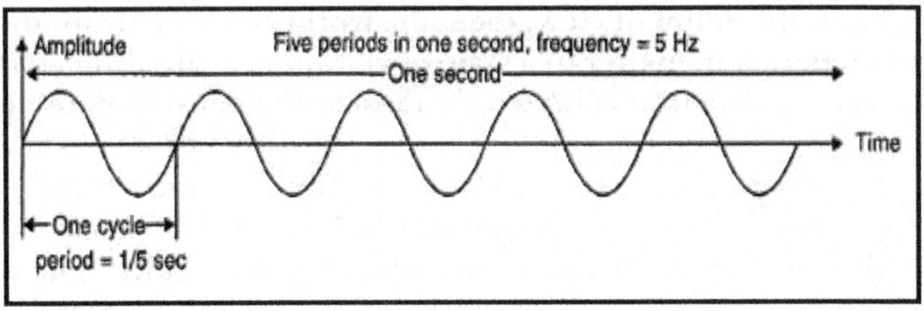

The greater the amplitude of a wave, the higher the level of energy it carries. With a sound wave, this means that the sound will be louder. With light, the bigger amplitude means the light will be brighter. In water, the bigger the amplitude of the waves, the higher they will be. Frequency is the number of occurrences of a repeating event per unit of time, so the higher the frequency, the more there is happening in each moment.

All of these characteristics of energy fall under the laws of physics, defined as the study of matter and its motion through space-time, and its related concepts of energy and force. More broadly, physics is the general analysis of nature conducted in order to understand how the universe behaves.

In order to navigate the physical world our senses take in a multitude of stimuli, much of it through our eyes and ears. What our senses take in, our minds process continuously, reinterpreting the incoming data from moment to moment, instantaneously making adjustments to the shifting environment. By regulating our breathing, heartbeat, and blood pressure, as well as other unconscious physiological functions, our body's natural intelligence strives for balance; especially when it is involved in intense physical activity that requires great coordination, like sports, dancing, and other performing arts.

This rapid exchange between inner subjectivity and outer objectivity is both an infinity pattern between external stimulus and inner interpretation, and by the nature of its interaction, an interference pattern. The continuous volley and serve of stimulus and response moves back and forth like a tennis match played at very high speed with multiple volleys.

Not only are these extremes doing a rapid subjective objective dance; the two sides of the brain are also communicating on many levels, instantaneously sending masses of information back and forth across the "net" of the corpus callosum, the central area in the fissure of the brain that facilitates communication between both sides.

These multichannel multifunctional energy exchanges happen everywhere throughout our bodies and minds down to the cellular and atomic level in everything from neurotransmitters to cellular respiration, the process that living organisms use to turn food into the energy they need for survival.

All of these manifestations of energy move directly from one pole to the other as direct current, or alternating back and forth in oscillations that constitute vibration about an equilibrium point characterized as a zero point, centered between positive and negative poles.

A STUDY IN PERCEPTION

At the center of experience the individual defines what they perceive of as real at the interference pattern that occurs between objectivity and subjectivity. Paying increased attention there cultivates what is called witness consciousness which brings greater meaning of the expression, "Where your attention goes, there your energy goes."

Witness consciousness becomes a heightened focal point that harnesses the energy of awareness that is responsible for the depth of our perception by the act of paying attention and observing. The Zen concept of non-attachment provides a good example of witness consciousness which is characterized as a practice of presence and mindfulness without allowing our sense of well being to rely upon anything other than our own presence of awareness. It means to be in the world, but not of the world.

Gurdjieff characterized witness consciousness this way.

"Instead of the discordant and often contradictory activity of different desires, there is *one single I*, whole, indivisible, and permanent; there is individuality, dominating the physical body and its desires and able to overcome both its reluctance and its resistance. Instead of the mechanical process of thinking there is *consciousness*. And there is *will*, that is, a power, not merely composed of various often contradictory desires belonging to different 'I's', but issuing from consciousness and governed by individuality or a single and permanent I. Only such a will can be called 'free', for it is independent of accident and cannot be altered or directed from without."

Our five primary mechanisms of perception come from our sense receptors; taste, sight, touch, smell, and hearing. With the exception of our sense of touch, which comes to us through all parts of our bodies, our other four senses come through our head, which filters and puts them together into the unique perspective that we as individuals experience.

Aside from the subject/object ground zero that puts the center of our universe between our eyes where we decide what our reality consists of according to our interpretation of these impressions, this location is the most logical place to locate it based on the construction of our body and the way our senses are arrayed about our head.

This focus of awareness whether physical, mental, or metaphysical, points to the notion of the third eye, the mind's eye, or inner eye that represents a mystical esoteric concept that refers to a speculative invisible eye reputed to provide perception beyond ordinary sight. This third eye is thought to be the extension of what the mind perceives in the form of a subconscious awareness of the surroundings and interactions of the environment. In some spiritual traditions the third eye refers to the gate that leads to inner realms and spaces of higher consciousness, and it often symbolizes a state of enlightenment or the evocation of mental images having deep personal, spiritual, or psychological significance. Some teachings view the concept of the third eye as a metaphor for non-dualistic thinking; the way mystics see.

The rudiments of a biological basis for the mind's eye is found in the deeper portions of the brain below the neocortex *where the center of perception exists.* The neocortex is characterized as a sophisticated memory storage warehouse where data received as input from sensory systems is compartmentalized via the cerebral cortex which allows shapes to be identified. Given the lack of filtering input produced internally, we have the ability to hallucinate and see things that aren't received as external input, but as internal. Not all people have the same internal perceptual ability. For many, when their eyes are closed, the perception of darkness prevails, while others can perceive colorful, dynamic imagery.

In Theosophy the third eye relates to the pineal gland. According to this theory, in far ancient times humans had an actual third eye in the back of the head with a physical and spiritual function. As humans evolved this eye atrophied and sank into what today is known as the pineal gland, a small endocrine gland in the vertebrate brain with a shape that resembles a pine cone, hence its name. It is located near the center of the brain, between the two hemispheres, tucked in a groove where the two halves of the thalamus join. From the point of view of biological evolution, the pineal gland represents a kind of atrophied photoreceptor, and in the epithalamus of some species of amphibians and reptiles it is linked to a light-sensing organ known as the parietal eye, which is also called the pineal eye or *third eye.* Philosopher René Descartes believed the pineal gland to be the "principal seat of the soul."

Phenomenology is the Western philosophical tradition that most forcefully called into question the modern assumption of a single, wholly determinable, objective reality. It has its source in Descartes' well-known separation of the thinking mind or subject from the material world of things or objects. This philosophy formed the basis for the divide-and-conquer western scientific method which has shown us many things, but falls short in comprehending the infinite vastness of reality. Instead of showing us more, our divide and conquer mentality has resulted in isolating us through technology and civilization in a divide that has grown by greater and greater degrees.

In terms of this growing separation, French phenomenologist Maurice Merleau-Ponty stated:

"All my knowledge of the world, even my scientific knowledge, is gained from my own particular point of view, or from some experience of the world without which the symbols of science would be meaningless. The whole universe of science is built upon the world as directly experienced, and if we want to subject science itself to rigorous scrutiny and arrive at a precise assessment of its meaning and scope, we must begin by reawakening the basic experience of the world, of which science is the second - order expression... To return to things themselves is to return to that world which precedes knowledge, of which knowledge always *speaks*, and in relation to which every scientific schematization is an abstract and derivative sign-language, as is geography in relation to the countryside in which we have learnt beforehand what a forest, a prairie, or a river is."

Regardless of our conception of the third eye or the mind's eye, whether physical, mental, or metaphysical, we cannot disregard the fact that the primary focus of our awareness and the creation of reality as we know it lies in our subjective interpretation of a world that exists through us and around us.

Aside from these physical, mental, and subjective indicators of the location Descartes refers to as the seat of the soul, for the more scientific minded there are objective indicators evident in physics, the branch of science that studies the nature and properties of matter and energy, including mechanics, heat, light and other radiation along with sound, electricity, magnetism, and the structure of atoms.

This phenomenon is known as the observer effect, which is the fact that simply observing a situation necessarily changes it. Physicists have discovered that even passive observation of quantum phenomena can in fact change it.

Gurdjieff taught that man consists of two parts: *essence* and *personality*. Essence in man is what is *his own*, which can be considered as coming from Source, and personality in man is what is "not his own", meaning what has come from outside, what he has learned, or *reflects*. This includes all traces of exterior impressions left in the memory and sensations, all words and movements that have been learned, and all feelings created by imitation. All that is "not his own", are subpersonalities created partly by the intentional influences of other people, and partly by voluntary and involuntary reflection of them as a child.

In Gurdjieff's words:

"Essence is the truth in man; personality is the false. But in proportion as personality grows, essence manifests itself more and more rarely and more and more feebly and it very often happens that essence stops in its growth at a very early age and grows no further. It happens very often that the essence of a grown up man, even that of a very intellectual and, in the accepted meaning of the word, highly 'educated' man, stops on the level of a child of five or six. This means that everything we see in this man is in reality 'not his own'. What is his own in man, that is, his essence, is usually only manifested in his instincts and in his simplest emotions. There are cases, however, when a man's essence grows in parallel with his personality. Such cases represent very rare exceptions especially in the circumstances of cultured life. Essence has more chances of development in men who live near to nature in difficult conditions of constant struggle and danger."

Like pieces of a broken mirror, or the proverbial drop of water in the ocean, each subpersonality contains the whole within it following the implicate/explicate order characteristic of holograms.

HOW WE PERCEIVE THE ENERGIES THAT CONSTITUTE REALITY

In a mechanical medium the energy of sound vibration is perceived in the motion of a tuning fork, a reed in a woodwind instrument, a harmonica, or the cone of a loudspeaker. In the audio spectrum, sound or pressure waves are generated by these vibrating structures, and in our bodies by our vocal cords, which are our main transmitting instruments of communication. These pressure waves induce vibration in the structures of our ear drums, our primary mode of receiving communication. Human hearing range falls within 20 to 20,000 Hz and we have two ears, one on each side of our head, allowing us to get a relative 360 degree input from this sense to alert us to what is going on around us, even in the dark.

Moving closer to the center of our head, our two eyes are set below our forehead directed in front of us the same way a predator's eyes are. We perceive in the portion of the electromagnetic spectrum visible to the human eye. Electromagnetic radiation in this range is called visible light or simply light. A typical human eye will respond to wavelengths from about 390 to 700 nm. In terms of frequency, this corresponds to a band in the vicinity of 430–770 THz.

This spectrum does not contain all the colors that the human eye and brain can distinguish. Unsaturated colors like pink or purple variations like magenta are absent because they can be made only by a mix of multiple wavelengths. Colors containing only one wavelength are called pure or spectral colors.

Visible wavelengths pass through the "optical window", the region of the electromagnetic spectrum that allows wavelengths to pass largely unattenuated through the Earth's atmosphere. An example of this phenomenon is clean air that scatters blue light more than red wavelengths making the midday sky appear blue. The optical window is also referred to as the "visible window" because it overlaps the human visible response spectrum.

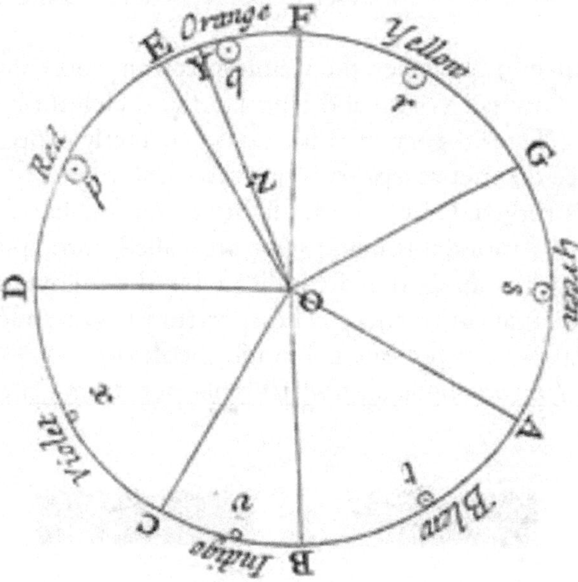

Newton's Color Circle

Newton's color circle shows the colors he associated with musical notes which might help explain the phenomenon of synesthesia where subjects hear color and see sound. The spectral colors from red to violet are divided by the notes of the musical scale, starting at D. The circle completes a full octave, from D to D. Newton's circle places red at one end of the spectrum next to violet at the other reflecting the fact that non-spectral purple colors are observed when red and violet light are mixed.

Newton divided the visual color spectrum into seven named colors: red, orange, yellow, green, blue, indigo, and violet. He chose the octave of seven colors out of a belief derived from the ancient Greek sophists of a connection between the colors, musical notes, the known objects in the solar system, and the days of the week.

The human eye is relatively insensitive to indigo's frequencies, and some people who have otherwise good vision cannot distinguish indigo from blue and violet. For this reason it was later suggested that indigo should not be regarded as a color in its own right, but as a shade of blue or violet. Evidence indicates that what Newton meant by indigo and blue does not correspond to the modern meanings of those color words. Comparing Newton's observation of prismatic colors to a color image of the visible light spectrum shows that indigo corresponds to what is today called blue, whereas blue corresponds to cyan.

The connection between the visible spectrum and color vision was explored by Thomas Young and Hermann von Helmholtz in the early 19th century. Their theory of color vision correctly proposed that the eye uses three distinct receptors to perceive color.

Colors produced by visible light within a narrow band of wavelengths of monochromatic light are called pure spectral colors. The various color ranges that are indicated in the following illustration are an approximation of these. The spectrum is continuous, with no clear boundaries between one color and the next.

https://en.wikipedia.org/wiki/Visible_spectrum - cite_note-18

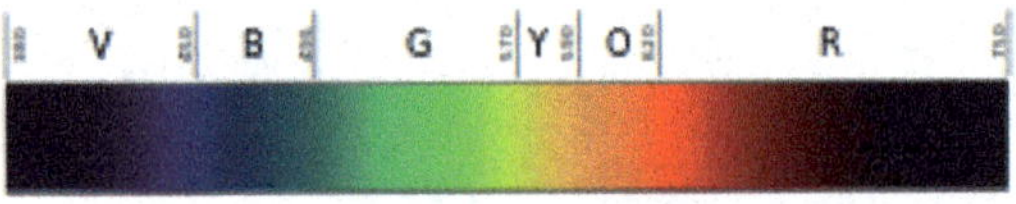

Color	Wavelength	Frequency	Photon energy
violet	380–450 nm	668–789 THz	2.75–3.26 **eV**
blue	450–495 nm	606–668 THz	2.50–2.75 eV
green	495–570 nm	526–606 THz	2.17–2.50 eV
yellow	570–590 nm	508–526 THz	2.10–2.17 eV
orange	590–620 nm	484–508 THz	2.00–2.10 eV
red	620–750 nm	400–484 THz	1.65–2.00 eV

Spectral Colors

Moving further inward, in the center of our forward-facing sensory organs is our protruding nose, which along with our mouth allows us to breathe, as well as functioning as our organ of smell. This organ is the site of specialized sensory receptor cells that transduce chemical substances and generate biological signals that can be in the form of an action potential if the chemoreceptor is a neuron, or in the form of a neurotransmitter that can activate a nearby nerve fiber if the chemosensor is a specialized sensory receptor cell like the taste receptor in a taste bud that forms the sense of smell.

Olfaction has many purposes, like the detection of hazards, pheromones, and food, and it integrates with other senses to form the sense of flavor. Olfaction occurs when odorants bind to specific sites on olfactory receptors located in the nasal cavity. Glomeruli aggregate signals from these receptors and transmit them to the olfactory bulb where the sensory input interacts with parts of the brain responsible for smell identification, memory, and emotion.

The process that codes olfactory information in the brain for perception is not fully understood. When an odorant is detected by receptors they essentially break the odorant down, and the brain puts it back together for identification and perception.

The average human is capable of distinguishing over one trillion unique odors. Researchers tested the psychophysical responses to combinations of over 128 unique odor molecules with combinations composed of up to 30 different component molecules and noted that this estimate is conservative. Some subjects of their research might be capable of deciphering between a thousand trillion odorants, adding that their worst performer could probably still distinguish between 80 million scents. The human olfactory system with its hundreds of different olfactory receptors far outperforms the other senses in the number of physically different stimuli it can discriminate.

Flavor perception is an aggregation of auditory, taste, haptic, and smell sensory information. Haptic perception is the ability "to grasp something." Perception in this case is achieved through the active exploration of surfaces and objects by a moving subject, as opposed to passive contact by a static subject during tactile perception.

During the process of mastication, the tongue manipulates food to release odorants that enter the nasal cavity during exhalation. The olfaction of food has the sensation of being in the mouth because of co-activation of the motor cortex and olfactory epithelium during mastication.

Humans have far fewer active odor receptor genes than other primates and other mammals. In mammals each olfactory receptor **neuron** expresses only one functional odor receptor. Odor receptor nerve cells function like a key–lock system: if the airborne molecules of a certain chemical fit into the lock, the nerve cell responds. There are a number of competing theories regarding the mechanism of odor coding and perception, but according to the shape theory, each receptor detects a feature of the odor molecule.

The weak-shape theory known as the odotope theory suggests that different receptors detect only small pieces of molecules, and these minimal inputs are combined to form a larger olfactory perception similar to the way visual perception is built up of smaller, information-poor sensations combined and refined to create a detailed overall perception. Researchers have found that a functional relationship exists between molecular volume of odorants and the olfactory neural response.https://en.wikipedia.org/wiki/Olfaction - cite_note-10.1101.2F013516-46

The vibration theory proposed by Luca Turin posits that odor receptors detect the frequencies of vibrations of odor molecules in the infrared range by quantum tunnelling. Although there is no theory that explains olfactory perception completely, the fact that seeing, hearing, and touch are frequency dependent lends weight to the vibration theory.

All of our senses, including our sense of touch ultimately bring us to the mouth, our organ of sustenance located below and intimately connected to the nose, which is directed by retronasal smell, retronasal olfaction, or mouth smell, giving us the ability to perceive the flavor dimensions of foods and drinks. Retronasal smell is a sensory modality that produces flavor best described as a combination of traditional, or orthonasal smell and taste modalities that create flavor from smell molecules in foods or drinks shunting up through the nasal passages while chewing. When people use the term smell, they are usually referring to orthonasal smell, or the perception of smell molecules that enter through the nose and up the nasal passages. Retronasal smell is critical for experiencing the flavor of foods and drinks. Flavor should be contrasted with taste, which refers to five specific dimensions: (1) sweet, (2) salty, (3) bitter, (4) sour, and (5) umami. Perceiving anything beyond these five dimensions, like distinguishing the flavor of an apple from a pear requires retronasal smell.

Taste in the gustatory system allows humans to distinguish

between safe and harmful food, and to gauge nutritional value. Digestive enzymes in saliva dissolve food into base chemicals that are washed over the papillae and detected as tastes by the taste buds. The tongue is covered with thousands of small bumps called papillae which are visible to the naked eye. Within each papilla are hundreds of taste buds. The exception to this are the filiform papillae that do not contain taste buds. There are between 2000 and 5000 taste buds located on the back and front of the tongue. Others are located on the roof, sides, the back of the mouth, and in the throat. Each taste bud contains 50 to 100 taste receptor cells.

Taste is a form of chemoreception that occurs in the five different types of specialized taste receptors in the mouth. Each receptor has a different manner of sensory transduction that detects the presence of a certain compound and starts an action potential that alerts the brain.

Where the head is the center of perception for four of our senses, the fifth sense is centered in the body, our primary organ of feeling, and at the center of that is our heart which is considered the center of feeling and emotion.

Many ancient cultures saw the heart as the true center of their personal universe, functioning as the portal connecting them through the sun of our local solar system through multiple permutations in a holographic manner all the way back to the infinite source of the great mystery.

Continuing our inward-bound tour to our center brings us to our tactile senses that literally cover our whole body with greater concentrations of nerve endings located at critical points like our fingertips and face. This multifaceted pervasive network is a complex system of sensory neurons and pathways known as our somatosensory system that responds to changes at the surface or inside the body. The axon nerve fibers of sensory neurons connect and respond to various sensory receptor cells that are activated by different stimuli like heat and pain, giving a functional name to the responding sensory neurons like thermoreceptors that carry information about temperature changes, making them vibratory sensors in that heat is a form of energy produced when molecules increase their rate of oscillation.

Other types of cells include mechanoreceptors, chemoreceptors, and nociceptors that send signals along a sensory nerve to the spinal cord where they are processed by other sensory neurons and relayed to the brain for further processing. Sensory receptors are found all over the body including the skin, epithelial tissues, muscles, bones, joints, internal organs, and the cardiovascular system.

Somatic senses are sometimes referred to as somesthetic senses with the understanding that somesthesis includes the sense of touch, proprioception, which is the sense of position and movement, and depending on usage haptic perception, which is related to the concept of extended physiological proprioception. When a tool like a stick is used, perceptual experience is transferred to the end of the tool. In this case haptic perception relies on the forces experienced during touch. Four mechanoreceptors in the skin each respond to different stimuli for short or long periods.

Merkel cell nerve endings are found in the basal epidermis and hair follicles. They react to low vibrations from 5–15 Hz and deep static touch like shapes and edges. Due to a small receptive field of extremely detailed information they are used in areas like fingertips the most. They are not covered and respond to pressures over long periods.

Tactile corpuscles react to moderate vibration (10–50 Hz) and light touch and are located in the dermal papillae. Due to their reactivity they are primarily located in fingertips and lips and respond in quick action potentials unlike Merkel. They are responsible for the ability to read Braille and feel gentle stimuli.

Lamellar corpuscles determine gross touch and distinguish rough and soft substances. They react in quick action potentials, especially to vibrations around 250 Hz up to centimeters away. They are the most sensitive to vibrations and have large receptor fields. Pacinian corpuscles only react to sudden stimuli, so pressures like clothes that always compress their shape are quickly ignored.

Bulbous corpuscles react slowly and respond to sustained skin stretch, and are responsible for the feeling of object slippage and play a major role in the kinesthetic sense and control of finger position and movement. Merkel and slow-response bulbous cells are myelinated while the other fast-response cells are not. All of these receptors are activated by pressures that squish their shape causing an action potential.

The receptor for the sense of balance resides in the vestibular system in the ear for the three-dimensional orientation of the head, and by inference, the rest of the body. Balance is also mediated by the kinesthetic reflex fed by proprioception that senses the relative location of the rest of the body to the head. In addition, proprioception estimates the location of objects which are sensed by the visual system providing confirmation of the place of those objects relative to the body as input to the mechanical reflexes of the body.

The somatosensory cortex encodes incoming sensory information from receptors all over the body. Affective touch is a type of sensory information that elicits an emotional reaction and is usually social in nature like physical human touch. This type of information is coded differently than other sensory information. Intensity of affective touch is still encoded in the primary somatosensory cortex, but the feeling of pleasantness associated with affective touch activates the anterior cingulate cortex more than the primary somatosensory cortex. Functional magnetic resonance imaging data shows that increased blood oxygen level contrast signals in the anterior cingulate cortex as well as the prefrontal cortex are highly correlated with pleasantness scores of an affective touch. Inhibitory transcranial magnetic stimulation of the primary somatosensory cortex inhibits the perception of affective touch intensity, but not affective touch pleasantness.

THE SYMPHONY OF INTERFERENCE PATTERNS WITHIN

Neural oscillations are rhythmic brainwaves of repetitive neural activity that occur in the central nervous system. Neural tissue can generate oscillatory activity in many ways, driven by mechanisms within individual neurons or by interactions between neurons. In individual neurons, waveforms can appear as oscillations in membrane potential or as rhythmic patterns of action potentials that activate post-synaptic neurons.

At the level of neural ensembles, synchronized activity of large numbers of neurons can give rise to macroscopic oscillations that can be observed in an electroencephalogram. Oscillatory activity in groups of neurons generally arise from feedback connections between neurons that result in the synchronization of their firing patterns. The interaction between neurons can also engender oscillations at a different frequency than the firing frequency of individual neurons.

Some possible roles of neural brainwaves include feature binding, information transfer mechanisms, and the generation of rhythmic motor output. Oscillatory activity in the brain is widely observed at different levels of organization and is thought to play a key role in processing neural information.

Neural oscillations are observed throughout the central nervous system at all levels, and include spike trains, local field potentials, and large-scale oscillations characterized by their frequency, amplitude, and phase. In large-scale oscillations, amplitude changes come from changes in synchronization within a neural ensemble, also referred to as local synchronization.

In addition to local synchronization, oscillatory activity of distant neural structures can synchronize and have been linked to many cognitive functions like information transfer, perception, motor control, and memory.

The first discovered, best-known frequency band is alpha activity within the 7.5–12.5 Hz range. It can be detected from the occipital lobe during relaxed wakefulness and increases when the eyes are closed.

Other recognized frequency bands are: delta, 1–4 Hz, theta, 4–8 Hz, beta, 13–30 Hz, and gamma, 30–70 Hz. Faster rhythms like gamma activity have been linked to cognitive processing. EEG signals change dramatically during sleep and show a transition from faster frequencies to increasingly slower frequencies like alpha waves. Different sleep stages are characterized by their spectral content, and neural oscillations have been linked to cognitive states like awareness and consciousness.

Neurons can generate rhythmic patterns of action potentials or spikes, and some types of neurons have the tendency to fire at particular frequencies referred to as *resonators*. When vibrations from one object match the resonant frequency of another object, the two are said to resonate because the first object amplifies the vibrations of the second object. Bursting is another form of rhythmic spiking. Spiking patterns are considered fundamental for information coding in the brain. If numerous neurons spike in synchrony, they can give rise to oscillations in local field potentials.

Neural oscillations studied from a mathematical framework belong to the field of neurodynamics, an area of research in the cognitive sciences that places strong focus on the dynamic character of neural activity in describing brain function. Neurodynamics considers the brain to be a dynamical system and uses differential equations to describe how neural activity evolves over time. In particular, it aims to relate dynamic patterns of brain activity to cognitive functions like perception and memory. The functions of neural oscillations are wide-ranging and vary for different types of oscillatory activity like the generation of rhythmic activity in a heartbeat and the neural binding of sensory features in perception, like the shape and color of objects.

Oscillatory activity is observed throughout the central nervous system at all levels of organization. Three different levels have been widely recognized: the micro-scale which is the activity of a single neuron, the meso-scale, which is the activity of a local group of neurons, and the macro-scale, which is the activity of different brain regions.

Neurons generate action potentials resulting from changes in the electric membrane potential and can generate multiple action potentials in sequence forming spike trains which are the basis for neural coding and information transfer in the brain. Spike trains can form all kinds of patterns, like rhythmic spiking and bursting, and often display oscillatory activity.

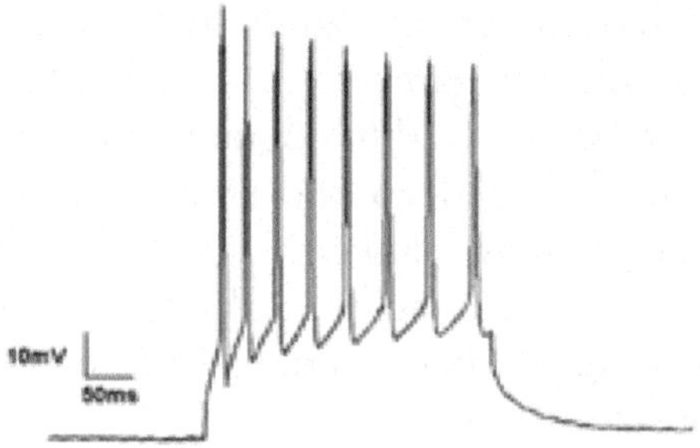

Tonic firing pattern of a single neuron showing rhythmic spiking activity

Neural ensembles can generate oscillatory activity https://en.wikipedia.org/wiki/Endogenousthrough local interactions between excitatory and inhibitory neurons. Inhibitory interneurons play an important role in producing neural ensemble synchrony by generating a narrow window for effective excitation and rhythmically modulating the firing rate of excitatory neurons.

Neural oscillation also arise from interactions between different brain areas coupled through the structural connectome. Because all brain areas are bidirectionally coupled, these connections between brain areas form feedback loops. Positive feedback loops cause oscillatory activity where frequency is inversely related to the delay time. An example of this kind of feedback loop is the connections between the thalamus and cortex. This thalamocortical network can generate oscillatory activity known as recurrent thalamo-cortical resonance. The thalamocortical network plays an important role in the generation of alpha activity.

In a whole-brain network model with realistic anatomical connectivity and propagation delays between 90 brain areas, oscillations in the beta frequency range emerge from the partial synchronization of subsets of brain areas oscillating in the gamma-band generated at the mesoscopic level. This occurs because synchronization in the presence of time delays reduces the oscillation frequency.

In addition to periodic spiking, subthreshold membrane potential oscillations like resonance behavior that does not result in action potentials, can also contribute to oscillatory activity by facilitating synchronous activity of neighboring neurons. Like pacemaker neurons in central pattern generators, subtypes of cortical cells fire bursts of spikes rhythmically at preferred frequencies. Bursting neurons have the potential to serve as pacemakers for synchronous network oscillations, and bursts of spikes can underlie or enhance neuronal resonance.

If a group of neurons engage in synchronized oscillatory activity, the neural ensemble can be mathematically represented as a single oscillator. Different neural ensembles are coupled through long-range connections and form a network of weakly coupled oscillators at the next spatial scale which can generate a range of dynamics, including oscillatory activity.

Long-range connections between different brain structures like the thalamus and the cortex involve time-delays due to the finite conduction velocity of axons. Because most connections are reciprocal they form feed-back loops that support brainwave activity. Oscillations recorded from multiple cortical areas can become synchronized to form large scale brain networks resulting in an expanded expression. Coherent activity of large-scale brain activity can also form dynamic links between brain areas required for the integration of distributed information.

Oscillations can be described and analyzed using mathematics which have identified several dynamical mechanisms that generate rhythmicity. Among the most important are linear harmonic oscillators, limit cycle oscillators, and delayed-feedback oscillators. Harmonic oscillations appear frequently in nature in the form of sound waves, the motion of a pendulum, and other vibrations and generally arise when a physical system is perturbed by a small degree from a minimum-energy state.

Linear oscillators and limit-cycle oscillators differ in how they respond to fluctuations in input. In a linear oscillator the frequency is relatively constant but the amplitude can vary. In a limit-cycle oscillator the amplitude tends to be constant, but the frequency can vary. A heartbeat is an example of a limit-cycle oscillation in that the frequency of beats varies widely, while each individual beat continues to pump about the same amount of blood.

Spontaneous activity is brain activity in the absence of an explicit task like sensory input or motor output, and is referred to as resting-state activity as opposed to activity induced by sensory stimuli or motor responses. Ongoing brain activity also has an important role in perception, as it interacts with activity related to incoming stimuli. EEG studies suggest that visual perception is dependent on both the phase and amplitude of cortical oscillations like what happens when the amplitude and phase of alpha activity at the moment of visual stimulation predicts whether a weak stimulus will be perceived by the subject.

In response to input, a neuron or neuronal ensemble can change the frequency it oscillates at, changing the rate that it spikes at. Often, a neuron's firing rate depends on the summed activity it receives. Frequency changes are also observed in central pattern generators and relate to the speed of motor activities like step frequency in walking, but changes in relative oscillation frequency between different brain areas is not so common because the frequency of oscillatory activity is typically related to time delays between brain areas.

Increases in oscillatory activity are referred to as event-related synchronization while decreases are referred to as event-related desynchronization. Neural synchronization can be modulated by task constraints like attention, and is thought to play a role in feature binding, neuronal communication, and motor coordination.

Synchronized firing of neurons also forms the basis of periodic motor commands for rhythmic movements produced by a group of interacting neurons that form a network called a central pattern generator. Central pattern generators are neuronal circuits that when activated can produce rhythmic motor patterns in the absence of sensory or descending inputs that carry specific timing information. Examples are walking, breathing, and swimming. Neuronal spiking is believed to be the basis for information transfer in the brain, and for such a transfer, information needs to be coded in a spiking pattern.

Oscillations at the spinal level become synchronized to beta oscillations in the motor cortex during constant muscle activation. Likewise, muscle activity of different muscles reveals inter-muscular coherence at multiple distinct frequencies reflecting the underlying neural circuitry involved in motor coordination, while neural oscillations, in particular theta activity, are extensively linked to memory function.

THE ARCHITECTURE Of A THREE POUND HOLOGRAPHIC UNIVERSE

Our physical bodies are bilateral with ten toes, ten fingers, and an opposable thumb, giving us a distinct physical advantage that puts us at the top of the earth's animal hierarchy with the higher primates. Physiologically our digits are a prime external indicator of our place at the top of the evolutionary scale, while internally the development of our brain charts the path of the evolution of our consciousness along its progression into self-awareness, the capability that distinguishes us above all other life forms on the planet.

The term consciousness refers to common capabilities of humans and other animals, and serves to differentiate between them, particularly when designating uniquely human linguistic, rational, and abstract capabilities. Consciousness also includes a range of functions that esoteric traditions claim supersede rational and egoic forms of consciousness, representing the evolution of what is conceptualized as spirit, soul, mind, self, and transcendental human capabilities.

Although human consciousness is not specific to any particular function or system of the brain, different modalities of consciousness are associated with different information-processing functions, integration of brain processes, and patterns of homeostasis. Every major system of the brain participates in complex human behavior, but specific systemic patterns of brain functioning are associated with distinct experiential states and modes of consciousness.

The relationship of brain physiology to consciousness is illustrated through an examination of how the physical structures of the brain and their associated activities relate to patterns of consciousness.

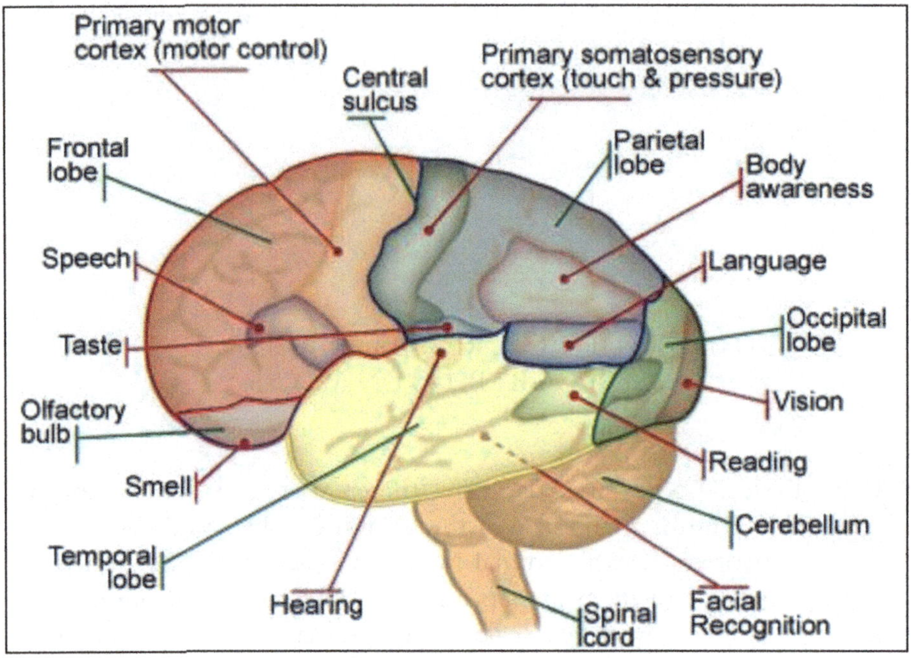

Diagram of Cerebrum Functions

The brain can be viewed as three anatomically distinct systems that are integrated to provide a range of behavioral, emotional, and informational functions. Human motor patterns, emotional states, and advanced cognitive and linguistic capabilities are managed by these systems that emerged sequentially in evolution. This triune brain model provides a framework for explicating the relationship between systemic brain activities and consciousness that relates to lower brain systems common with other animals and to unique aspects of the human brain.

The hierarchical tripartite brain is based on neuroanatomical, structural, and functional divisions that break down into three strata starting with the reptilian, followed by the paleomammalian, and neomammalian brains. These three formations have different anatomical structures that mediate different psychological and behavioral functions with their own forms of subjectivity, intelligence, time and space sense, memory capabilities, and motor functions. Although the three segments are integrated, they represent a functional hierarchy of information-processing capabilities that provide the basis for distinct forms of consciousness.

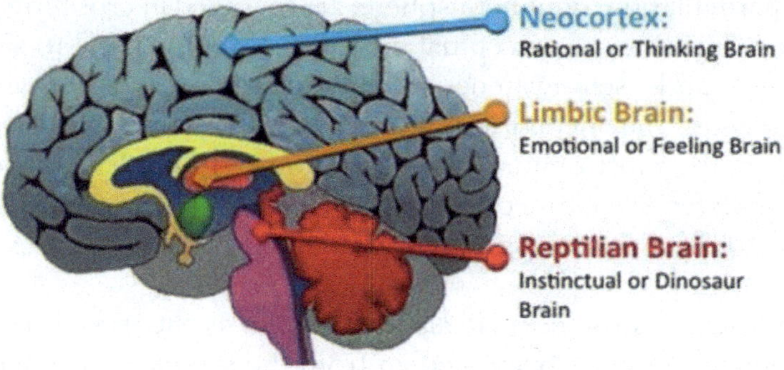

The Triune Brain

The reptilian brain is composed of the upper spinal cord, portions of the mesencephalon or midbrain, the diencephalon which contains the thalamus-hypothalamus, and the basal ganglia.

The reptilian brain regulates organic functions like metabolism, digestion, and respiration, and is responsible for wakefulness, attentional mechanisms, and the regulation and coordination of behavior.

The paleomammalian brain is based on evolutionary developments in the limbic system that makes a distinctions between reptiles and mammals. This structure provides the basis for social behavior, nonverbal, emotional, and analogical information processing, and it functions as an emotional brain, mediating affect, sex, fighting and self-defense, social relations, bonding and attachment, and the sense of self that provides the basis for beliefs, certainty, and convictions.

The neomammalian brain provides the basis for advanced symbolic processes, culture, language, logic, rational thought, analytical processes, and complex problem solving.

The reptilian brain provides the basic plots and actions of the body and the paleomammalian brain provides the emotional influences on thoughts and behavior while the neomammalian brain uses enhanced symbolic capacities to elaborate on the basic plots and emotions while integrating them with higher-level information processing.

The reptilian brain provides the organism with primary or simple awareness, which is adaptation to the environment through reflexes, conditioned responses, and habituation, as well as through instrumental learning. The paleomammalian brain provides for qualities of consciousness enriched by self, other, and emotions, while the neomammalian brain encompassing the tertiary neocortical area and in particular the right hemisphere, are involved in cross-integration and reorganization of perceptual modalities that are basic to symbolic cognition and self-awareness. The neomammalian brain that encompasses the neocortex and connecting thalamic structures represents the most dramatic evolution of the brain. Its functions are based on extensive connections with the visual, auditory, and somatic systems, indicating the primary orientation of the neocortex to the external world.

These three structures are also referred to as the three centers that we principally operate from, known from the top down as intellectual, emotional, and moving, forming a triad that relates to the every day functioning of our personality. These centers act as primary energizers and determine the way that we respond to stimuli.

A moving centered person tends to be physically active and fond of sports, travel, and action.

An emotionally centered person tends to be more perceptive and experiences situations in terms of likes and dislikes.

An intellectually centered person tends to be more verbal and enjoys philosophy and thinking for its own sake.

Most people react from one of these centers and when presented with any situation, their first response will be either thought, feeling, or action. If they react with thought they are intellectually centered, if they respond with feeling, they are emotionally centered, and if they respond with action, they are moving centered. People usually respond out of these three centers in a specific order based on their orientation.

If they are intellectually centered they first react with thought, emotions will follow, then they will act. If they are emotionally centered they will feel first, act, then think about it later. Any combination of the three is possible as a habitual pattern based on their most developed center which comes first, and their least developed which comes last. The closer a person comes to operating simultaneously out of all three centers, the more integrated and effective they are. When a person is fully aware and paying attention they are more likely to feel, think, and act simultaneously, and from this balanced point of power, fears are neutralized and experience shifts to insight, relatedness, and beauty. In order to reach higher levels of awareness defined as Maslow's peak experience, three conditions are necessary; a powerful desire to know the truth, a willingness to be emotionally open to life, and a practiced ability to be balanced energetically. Intellectual, emotional, and moving centers form a triad that relates to the universal building blocks of love, energy, and truth, otherwise known as, love, power, and wisdom, a triad that manifests as inspiration, action, and expression.

A natural result of the evolution of the human brain is the fragmentation of consciousness, reflecting both the increasing modularity of consciousness and the diversification of self. Shamanic traditions institutionalized procedures to overcome this fragmentation of consciousness by synchronizing this divergent human cognition through practices using altered states of consciousness to induce integrative brain processes. The use of external symbols and the relationships of the symbols to cultural psychodynamics engage transformative processes through entraining neurocognitive structures, provoking a restructuring of the self at levels below conceptual and operational thought.

Shamanistic practices induce extraordinary experiences and healing by producing integrative relationships among brain systems and psychocultural beliefs that reflect the simultaneous elicitation and integration of normal modes of information processing and consciousness that do not ordinarily occur together. A primary focus of the psychophysiological effects of shamanistic altered states is in the limbic system, or the paleomammalian brain. This brain area emerged in the evolution of mammals and provided a number of distinctive developments. Shamanistic healing is based on manipulation of processes and functions of the paleomammalian brain, including self-identity, social identity, and their attachments, emotions, meanings, and references. Shamanistic altered states elicit processes of the paleomammalian brain and induce a systemic integration of information processing functions across the functional layers of the brain, producing limbic-cortical integration and interhemispheric synchronization, best symbolized by an interference pattern manifesting in the form of a dynamic figure 8 infinity pattern referred to as the infinity zone.

Shamanic healing practices achieve this integration by physically stimulating systematic brain wave discharge patterns that activate affects, memories, attachments, and other psychodynamic processes of the paleomammalian brain. This activation forces normally unconscious or preconscious primary information processing functions and outputs to be integrated into the operations of the frontal cortex. This integrates implicit understandings, socioemotional dynamics, repressed memories, unresolved conflicts, intuitions, and nonverbal visual, mimetic, and presentational knowledge into self-conscious awareness.

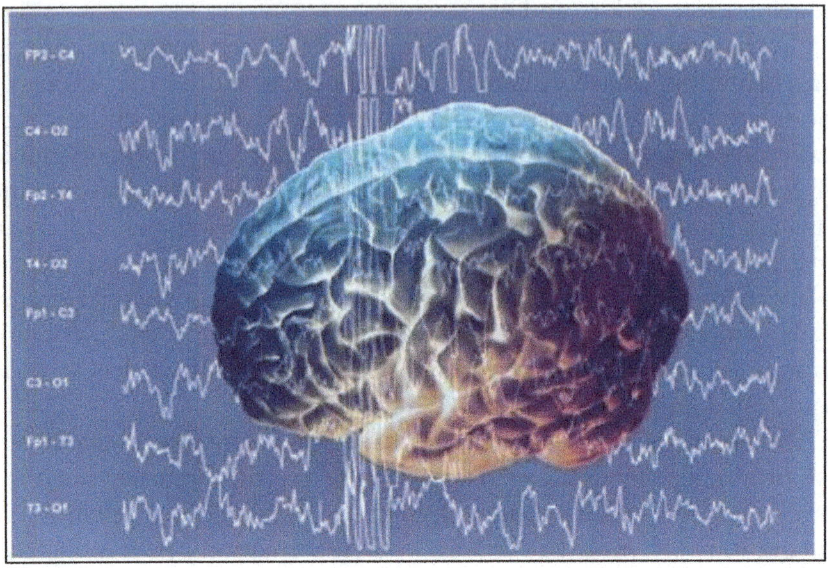

Activated Brain Wave Patterns

The desire to alter consciousness is an innate, human, biologically based drive with adaptive significance. The altered states of shamanism are a manifestation of a fundamental homeostatic dynamic of the nervous system. These manifestations of consciousness involve a biologically based integrative mode of consciousness, replacing normal waking conditions of sympathetic dominance and desynchronized fast wave activity of the frontal cortex with a parasympathetic dominant state. This heightened awareness is characterized by high voltage, slow-wave activity originating in the circuits linking the brain stem and the hippocampal-septal area of the limbic system with the frontal cortex. This high-voltage, slow wave activity originates in the hippocampal-septal area and imposes a synchronous slow-wave pattern on the frontal lobes, producing interhemispheric synchronization and coherence, limbic-cortex integration, and integration across the neuraxis, synthesizing behavior, emotion, and thought.

A primary characteristic of integrative consciousness involves the hierarchical integration of brain mechanisms, especially as manifested in limbic system driving of the frontal cortex through serotonergic induced integration across the neuraxis. This represents the integration of preconscious or unconscious functions and material into self-conscious awareness.

There are four different biologically based modes of human

consciousness; waking, sleeping, dreaming, and integrative consciousness. These reflect the fundamental aspects of systemic functioning of the human organism that meet the following system functions and needs, respectively; learning, adaptation, and survival needs in waking, recuperative functions, regeneration, and growth in deep sleep, memory integration and consolidation and psychosocial adaptation in dreaming, and psychodynamic growth and social and psychological integration.

The emotionally connected "female" right hemisphere of the brain reflects a different symbolic form of representation in dreaming than that used in waking consciousness. The bizarreness of dreams reflects the imaginative and creative capacities of the presentational modality. Dreams appear bizarre and illogical from the point of view of waking consciousness because dreams involve a different system of information representation, processing, and consolidation. This visual-spatial system of symbolic presentation is normally inhibited by the dominance of the "male" left-hemisphere verbal representational systems. Conditions that attenuate the left hemisphere's verbal representational systems allow expression of this presentational intelligence.

Meditation, rituals, hallucinogens and other shamanic practices can alter consciousness and physiological processes through a variety of mechanisms that induce retuning of the autonomic nervous system balance. This retuning tends to block the dominant hemisphere's functions and produces an integrative fusion with functions of the nondominant hemisphere. This structurally synchronized state tends to resolve internal conflicts and produce euphoric states. This synchronization depends on the elicitation of processes of lower brain structures that are associated with basic behavior, intentionality, and emotions.

These altered states evoke communicative responses from the paleomammalian brain that provide the basis for an expansion of consciousness by integrating information from the lower brain systems into operational activities of the frontal brain and establishing synchrony with the frontal brain that permits symbolic reprogramming of the emotional dynamics and behavioral repertoires of lower brain centers.

When the triune brain works in unison and the individual operates simultaneously out of all three centers, the more integrated and effective they are because they are fully aware. The instinctual, emotional, and intellectual centers will feel, think, and act together from a balanced point of power where fears are neutralized and experience shifts to insight, relatedness, and beauty.

When there is a homeostatic dynamic of the nervous system, a biologically based integrative mode of consciousness comes into play, replacing normal waking conditions – sympathetic dominance and desynchronized fast wave activity of the frontal cortex with a parasympathetic dominant state of interhemispheric synchronization, coherence, and integration across the neuraxis that synthesizes behavior, emotion, and thought. The parasympathetic slow-wave synchronization of the frontal cortex, and interhemispheric integration reflect activation of basic aspects of brain operation related to sensory and physiological integration, mental and emotional integration, and insight and transcendence.

In essence, the slow-wave synchronization of the frontal cortex is the energy that comes from transcending the male and female polarities, creating a triad that focuses on the energy manifest between the two as opposed to that of the extremes.

Three brains from the top down working in unison produce a fully aware state of insight, relatedness, and beauty, combined with the right "female" brain and the left "male" brain working in synchronization and coherence, synthesizing behavior, emotion, and thought to produce a slow-wave synchronization of the third frontal cortex indicate sensory and physiological integration, mental and emotional integration, and insight and transcendence.

Integration, Insight, and Transcendence

Male **Female**

Fully Integrated 3 x 3 Brain

If we are aware enough to be in harmony with the natural cosmic laws of oneness, reciprocity, and balance, and we are in harmony and alignment within ourselves, living in full integrity and unity with all that is, we will ride the wave, flow forward, and become fully conscious co-creators of our holographic reality, operating from our place of power at the crossing of the subjective and objective interference pattern of the infinity zone.

THE QUANTUM BRAIN

In the early 1940s, Karl Pribram, a neurophysiologist at Stanford University became interested in the concept of the brain as a hologram at a time when it was generally believed that each memory had a specific location somewhere in the brain's cells. These memory traces were called engrams. No one knew what they were made of, or whether they were neurons or special molecules. Most scientists at the time were confident that it was only a matter of time before one would be found because of research done by a Canadian neurosurgeon named Wilder Penfield in the 1920s that offered convincing evidence that specific memories had specific locations in the brain.

In a series of experiments while operating on the brains of epileptics, Penfield electrically stimulated different areas of their brain cells. To his amazement, when he stimulated the temporal lobes of one of his fully conscious patients, they reexperienced memories of past episodes from their lives in vivid detail.

In 1946, Pribram went to work with neuropsychologist Karl Lashley in Orange Park, Florida where Lashley was training rats to perform a variety of tasks like running mazes before removing different portions of their brains and retesting them. Pribram's aim was to cut out the area of the rat's brains containing the memory of their maze running abilities, but to his surprise he found that no matter what portion of their brain he cut out, he could not eradicate their memories. The rats' motor skills became impaired and they stumbled clumsily through the mazes, but even with massive portions of their brains removed, their memories remained intact.

From this observation, Pribram came to the conclusion that memories were not localized at specific brain sites, but were distributed throughout the brain. He knew of no mechanism or process to account for it, but the more he thought about it, the more convinced he became of the reality that it presented.

In the mid-sixties he read an article in *Scientific American* that described the construction of a hologram that provided a solution to the puzzle he had been wrestling with for so many years.

A hologram is created when a laser beam is split into two separate beams. The source beam is bounced off the object to be photographed like the apple in the illustration, and the second diffused beam collides with the reflected light of the first beam resulting in an interference pattern that is recorded on film.

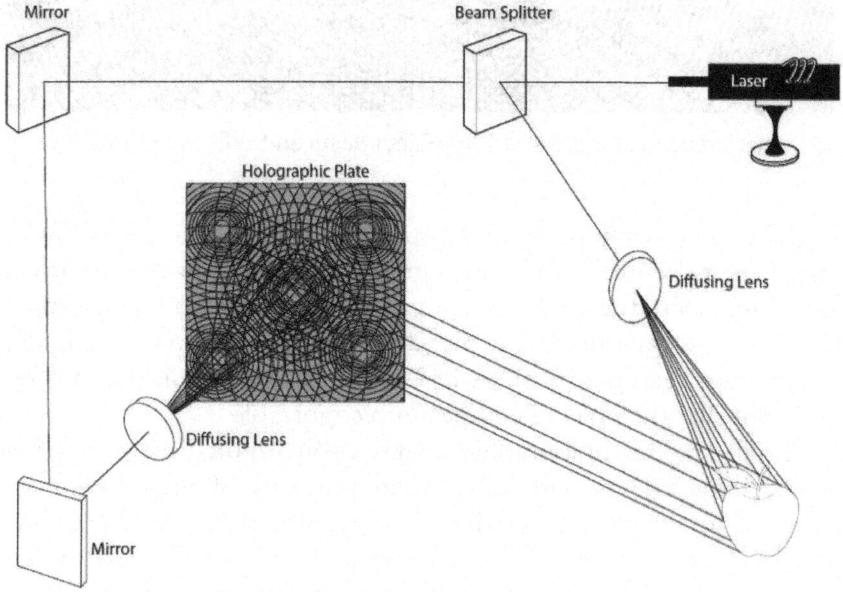

Interference is the criss-crossing pattern that occurs when two or more waves ripple through each other like dropping a pebble into a pond that produces a series of concentric waves that expand outward. If you drop two pebbles in a pond, you get two sets of expanding waves that pass through each other. The complex arrangements of crests and troughs that result from their collisions is known as an interference pattern.

The image on the film does not look like the object photographed, it looks like the image below, but as soon as another laser or a bright light shines through the film, an image of the original object reappears in three dimensions that can be viewed from different angles like it was the original object, but if you reach out and try to touch it, your hand will go through it.

Interference pattern caused by object beam and reference beam

Like the reflections in a broken mirror, if this image is cut in half, each half will contain the entire image, and if the halves are divided again and again, the entire image can still be reconstructed from each small piece of the film. Every piece of holographic film contains all the information recorded in the whole until the resolution diminishes to the point that the image becomes unrecognizable.

Similar to this holographic quality of distributed memory proven by the experiments with rats, when portions of their brains were removed in an effort to eradicate their memories, after removing as much as 90 percent of their visual cortex, they could still perform tasks requiring complex visual skills. In another experiment as much as 98 percent of a cat's optic nerves were severed without impairing their ability to perform complex visual tasks.

Electrical communications that take place between the brain's neurons do not occur alone. Neurons possess branches like tiny trees and when an electrical message reaches the end of one of these branches, it radiates outward like a ripple in a pond. Neurons are packed together so densely that these expanding ripples of electricity constantly criss-cross one another creating a kaleidoscopic symphony of interference patterns that gives the brain its holographic properties.

These seemingly infinite characteristics of holograms give them an amazing capacity for information storage. By changing the angle that two lasers strike a piece of photographic film, many different images can be recorded on the same surface. Any images recorded this way can be retrieved by illuminating the film with a laser beam possessing the same angle as the original two beams. Using this method, researchers have calculated that a one inch square of film can store the same amount of information contained in 50 Bibles.

In the creation of a hologram, the source beam is bounced off of a mirror and the second beam is allowed to collide with the reflected light of the first. The resulting interference pattern that is recorded comes from both reflections, one that is reflected off of the object and one that is reflected off of the mirror that creates the interference pattern.

Holograms have the properties of an infinite hall of mirrors, reflecting upon a reflection upon a reflection infinitely in a multitude of innumerable pieces that all contain the whole.

In the same way that the resolution in a hologram that is continuously broken down into smaller parts diminishes to the point that the image becomes unrecognizable, quantum physicists discovered that by breaking down matter itself into smaller and smaller pieces, you also reach a point where those pieces, like electrons, protons, and so on, no longer possess the traits of objects. Although an electron can act as if it were a miniscule dense particle, it possesses no dimension. It is impossible to measure the width of an electron because it is not an object as we know it.

An electron can manifest as a particle or a wave that can dissolve into a blurry cloud of energy and behave like a wave spread out over space. When an electron manifests as a wave it can do things particles cannot. If it is fired at a barrier that two slits have been cut into, it can go through both slits at the same time. When wavelike electrons collide with each other they create interference patterns, which are the basis of creating a hologram. The fact that an electron can manifest as either a particle or a wave is common to all subatomic particles as well as to all things once thought to manifest exclusively as waves like light, gamma rays, radio waves, and X-rays which can all change from waves to particles and back again.

Subatomic particles are not classified as waves or particles, but as a single category of something that is both called quanta, the plural of quantum. One electron is a quantum and several electrons are a group of quanta. The word quantum defines a wave particle, a term that refers to something that possesses both particle and wave qualities. The only time quanta manifest as particles is when we are looking at them. When an electron isn't being looked at it is always a wave.

One of the most fascinating paradoxes of the mystical world of quantum physics is the state of interconnectedness that exists between seemingly unrelated subatomic events. The Danish physicist Niels Bohr pointed out that if subatomic particles only come into existence in the presence of an observer, then it is meaningless to speak of a particle's properties and characteristics as existing before they are observed, which implies that subatomic particles are interconnected. This revelation led to the discovery that some subatomic processes result in the creation of a pair of particles with identical or closely related properties.

As an example, the positronium atom is composed of an electron and a positron which is an electron with a positive charge. Because a positron is the electron's antiparticle opposite, the two eventually annihilate each other and decay into two quanta of light or photons traveling in opposite directions. According to quantum physics no matter how far apart the photons travel, when they are measured they will always be found to have identical angles of polarization, which is the spatial orientation of the photons wavelike aspect as it travels away from its point of origin. This indicates that the two photons must be instantaneously communicating with each other in order to know which angle of polarization to agree upon.

Physicist David Bohm encountered another example of quantum interconnectedness while working with plasmas, which are gases containing a high density of electrons and positive ions manifesting as atoms with a positive charge. In his experiments Bohm discovered that once these atoms were in a plasma, like flocks of birds and schools of fish, electrons stopped behaving like individuals and started behaving as part of a larger, interconnected whole. Although their individual movements appeared random, vast numbers of electrons produced complex organized effects similar to a living organism. The plasma constantly regenerated itself and enclosed all impurities in a wall in the same way that a biological organism encased a foreign substance in a cyst.

In 1947 Bohm extended his research to the study of electrons in metals, and once again found that the seemingly haphazard movements of individual electrons produced highly organized overall effects. Like his plasma studies, these were no longer situations involving two particles each behaving as if it knew what the other was doing, but entire oceans of particles, each behaving as if it knew what untold trillions of others were doing.

This concept of quantum interconnectedness is referred to as nonlocality which Albert Einstein referred to as "spooky actions at a distance". It describes the ability of objects to instantaneously know about each other's state, even when separated by large distances, as if the universe at large instantaneously arranges its particles in anticipation of future events.

Nonlocality indicates that "separate" parts of the universe are connected in an intimate and immediate way. Nonlocality occurs due to the phenomenon of entanglement where particles that interact with each other become permanently correlated or dependent on each other's states and properties to the extent that they lose their individuality and behave as a single entity, a paradigm that also describes the interaction of the trillions of cells that comprise the human body.

The concepts of nonlocality and entanglement go hand in hand on multiple levels and are proven facts of quantum systems demonstrated in laboratory experiments.

If a pair of electrons are created together, one will have a clockwise spin and the other will have a counterclockwise spin. Quantum theory suggests a possible superposition where the two electrons can be considered to simultaneously have clockwise and counterclockwise spins. If the pair are separated by any distance and later checked, the second particle can be seen to instantaneously take the opposite spin to the first, so that the pair maintains its zero total spin, no matter how far apart they may be, essentially balancing each other out.

Implicit within this overarching paradigm are the concepts of implicate and explicate order for quantum theory, terms coined by Bohm to describe two different frameworks for understanding the same phenomenon or aspect of reality. These concepts were developed to explain the bizarre behaviors of subatomic particles that quantum physics describes and predicts with elegant precision.

Bohm used these ideas to describe how the appearance of these phenomena can appear differently, or how they can be characterized by varying factors, depending on contexts. The implicate, or "enfolded" order is seen as a deeper and more fundamental order of reality, while the explicate or "unfolded" order includes the abstractions that humans normally perceive.

In Bohm's words: "In the enfolded order, space and time are no longer the dominant factors determining the relationships of dependence or independence of different elements. Rather, an entirely different sort of basic connection of elements is possible, from which our ordinary notions of space and time, along with those of separately existent material particles, are abstracted as forms derived from the deeper order. These ordinary notions in fact appear in what is called the 'explicate' or 'unfolded' order, which is a special and distinguished form contained within the general totality of all the implicate orders."

The notion of implicate and explicate orders emphasizes the primacy of structure and process over individual objects. The latter are seen as approximations of an underlying process and in this approach, quantum particles and other objects are understood to have a limited degree of stability and autonomy.

Bohm believed that the weird behavior of quantum particles is caused by unobserved forces, maintaining that space and time might actually be derived from an even deeper level of objective reality. In the words of F. David Peat, Bohm considered that what we take for reality are "surface phenomena, explicate forms that have temporarily unfolded out of an underlying implicate order," meaning the implicate order is the ground that reality emerges from.

Bohm and other physicists worked toward a model of quantum physics where the implicate order is represented by an appropriate algebra or other pregeometry that considered spacetime as part of an explicate order connected to an implicate order they called *pre-space*. The spacetime manifold and the properties of locality and nonlocality all arise from an order in this pre-space. Others have suggested that an implicate order could be carried by an algebra, with the explicate order being contained in the various representations of this algebra.

Bohm emphasized the primary role of the implicate order's structure, stating: "My attitude is that the mathematics of the quantum theory deals *primarily* with the structure of the implicate pre-space and with how an explicate order of space and time emerges from it, rather than with movements of physical entities, such as particles and fields."

In another analogy, Bohm gave the example of a pattern produced by making small cuts in a folded piece of paper and then unfolding it. Widely separated elements of the pattern are, in actuality, produced by the same original cut in the folded piece of paper. Here, the cuts in the folded paper represent the implicate order, and the unfolded pattern represents the explicate order.

Bohm employed the hologram as a means of characterizing implicate order, noting that each region of a photographic plate where a hologram is observable contains within it the whole three-dimensional image that can be viewed from a range of perspectives.

According to Bohm, an electron is not one thing but a totality or ensemble enfolded throughout the whole of space. When an instrument detects the presence of a single electron it is because one aspect of the electron's ensemble has unfolded. When an electron appears to be moving it is due to a continuous series of dynamic unfoldments and enfoldments. Electrons and all other particles are no more permanent than the form a geyser of water takes as it gushes out of a fountain sustained by a constant flux from the implicate order. When a particle appears to be destroyed it is not lost, it has enfolded back into the deeper order that it sprang from.

Since it is made up of water, we can say that the ocean is contained in every drop. Similarly, a ray of the Sun contains the Sun itself, and in a seed, the entire structure of the tree is contained within it. Additionally, every cell of our body contains the complete information about the entire mind-body system. Each of the 100 billion cells that make up our bodies contain the complete version of the original DNA that was the source of our entire body.

A piece of holographic film and the image it generates are also examples of an implicate and explicate order. The film is an implicate order because the image encoded in its interference patterns is a hidden totality enfolded throughout the whole. The hologram projected from the film is an explicate order because it represents the unfolded and perceptible version of the image. The constant flowing exchange between the two orders explains how particles like the electron in the positronium atom can shift from one kind of particle to another. This shifting can be viewed as a particle, as an electron, folding back into the implicate order, while another, a photon, unfolds and takes its place which explains how a quantum can manifest as either a particle or a wave.

According to Bohm, both aspects are always enfolded in a quantum's ensemble, but the way an observer interacts with it determines which aspect unfolds and which remains hidden. The role an observer plays in determining the form a quantum takes may be analogous to the way a jeweler manipulates a gem to determine which of its facets become visible in which do not.

Because the term hologram usually refers to a static image, it does not convey the dynamic and ever active nature of the incalculable enfoldings and unfoldings that create our universe moment by moment. Bohm described the universe not as a hologram, but as a holomovement.

The existence of a deeper and holographically organized order also explains why reality becomes nonlocal at the subquantum level. When something is organized holographically, all semblance of location breaks down. The fact that every part of a piece of holographic film contains all the information possessed by the whole indicates that the information is distributed nonlocally, so if the universe is organized according to holographic principles, it also has nonlocal properties.

It is a known fact that space is filled with light and other electromagnetic waves that constantly crisscross and interfere with one another, and as has been firmly established by observation and endless experimentation, all particles are also waves, meaning physical objects and everything else we perceive are composed of interference patterns, a fact that validates Bohm's assertion that the universe is a holomovement.

When considered together Bohm and Pribram's theories provide a novel way of perceiving the great mystery that we exist within. Looking at it from this perspective, our brains mathematically construct objective reality by interpreting frequencies that are projections from another dimension which is a deeper order of existence beyond both space and time. Within this unfolding paradigm the brain itself is a hologram enfolded in a holographic universe.

All levels of creation are analogous to each other because each level of creation is an expression of the whole, interpreted through a different vibratory lens that constitutes a unique perspective, and every single thing within this interconnected reality is made of energy.

THE MATHEMATICAL IMPERATIVE

There are over 6500 spoken languages in the world, including written languages, local languages, and global languages. Each language has its own set of rules that include grammar, semantics, syntax, morphology, phonology, slang, colloquialisms, and any number of other nuances.

The quadrivium is the study of number and its relationship to space and time. Arithmetic is pure number, geometry is number in space, music is number in time, and astronomy is number in space and time. Regardless of the diversity of languages and all their nuances that have arisen throughout the ages in cultures all over the world, mathematics and all of its expressions that constitute the quadrivium are a universal language.

Music was originally the classical subject of harmonics, in particular the study of the proportions between the musical intervals created by the division of a monochord. A harmonic is a wave with a frequency that is a positive integer multiple of the fundamental frequency of the original periodic signal like a sinusoidal wave. The original signal is called the 1st harmonic, the other harmonics are known as higher harmonics. As all harmonics are periodic at the fundamental frequency, the sum of harmonics is also periodic at that frequency and the set of harmonics forms a harmonic series.

Harmonics are employed in various disciplines, including music, physics, acoustics, electronic power transmission, radio technology, and other fields. As an example, if the fundamental frequency is 50 Hz, a common AC power supply frequency, the frequencies of the first three higher harmonics are 100 Hz which is the 2nd harmonic, 150 Hz is the 3rd harmonic, and 200 Hz is the 4th harmonic. Any addition of waves with these frequencies is periodic at 50 Hz.

In music which contains notes, tones, melodies, and rhythms, harmonics are used on string instruments and wind instruments as a way of producing sound on the instrument, particularly to play higher notes.

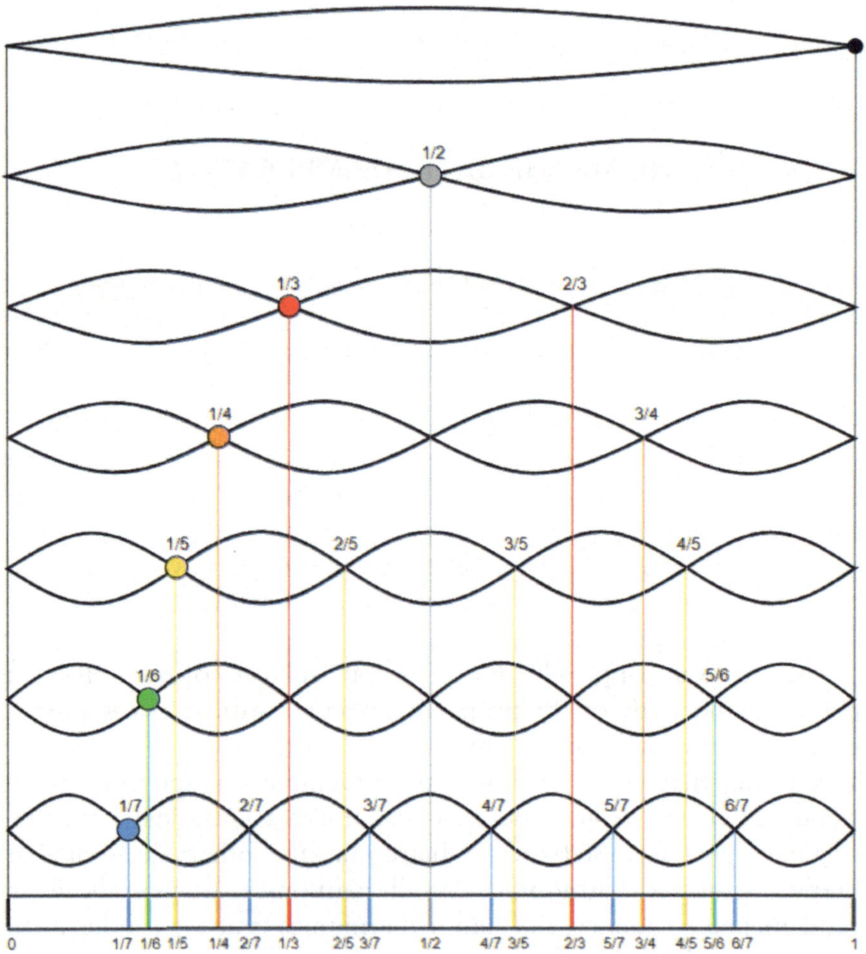

The nodes of a vibrating string are harmonics.

Within the arts and sciences that have evolved out of the quadrivium, everything is inspired from nature and nurtured in the imagination, and geometry is embedded everywhere in nature, while the gamut of shapes around us shapes our perspective.

The circle is particularly significant in its manifestation. The planets to stars to celestial bodies within our space are all spherical. There are other forms that add variety and beauty to natural formations like triangles, squares, pyramids, cones, cylinders, pentagrams, dodecahedrons and multiple complex expanding progressions beyond these basic forms.

Since geometry and its reciprocal mathematical ratios are evident in the human ecosystem, the human body, and in all life forms, it is the basis for understanding the natural and human-made universe. While used for sketching to calculating distances, the concepts are also used for creating animations, video games, music, architecture, art, weather patterns, and more.

There is far more to geometry than simple measuring and mapping.

While ancient texts describe a state of humanity where one language was spoken and referred to as a means of communication understood by all living things within specific cultural contexts, at its core is the need to comprehend a universal language understood by all.

Information is universal in all natural and artificial processes, and mathematical models can define all of them, making geometry a universal language due to its universal principles, foundations, and information processing in humans. All living beings, and machines are essential aspects of dynamic systems that are influenced by the interactive environment which includes the language of information in nature.

All living beings are information processing systems and all information is connected. Information can be interpreted in many different ways; as a message, the meaning of a message, or the effect of a message. It can also be the knowledge itself, a process in itself, or an entity in and of itself, and it can have a different meaning in different environments.

Many people think that mathematics is a human invention and to this way of thinking it is like a language. It can describe real things in the world but it doesn't "exist" outside the minds of the people who use it. The Pythagorean school of thought in ancient Greece believed that reality is fundamentally mathematical. From the shape of soap films, to gear design in engines, to the location and size of the gaps in the rings of Saturn, mathematics is everywhere.

Plato thought mathematics described objects that really exist, and these objects included numbers and geometric shapes. In modern times we can add more complicated mathematical objects like groups, categories, functions, fields, and rings. Plato also maintained that mathematical objects exist outside of space and time which deepens the mystery of how mathematics explains anything. This explanation involves showing how one thing in the world depends on another, but if mathematical objects exist in a realm apart from the world we live in, they don't seem capable of relating to anything physical.

The ancient Pythagoreans agreed with Plato that mathematics describes a world of objects, but they didn't think mathematical objects existed beyond space and time. They believed that physical reality is made of mathematical objects in the same way that matter is made of atoms.

Swedish-US cosmologist Max Tegmark and Australian physicist-philosopher Jane McDonnell have their own perspectives on this. Tegmark argues that reality is one big mathematical object like the simulation of a computer program, while McDonnell thinks reality is made of mathematical objects and minds, and that mathematics is how the Universe, which is conscious, comes to know itself.

Others believe that the world has two parts, mathematics and matter. Mathematics gives matter its form, and matter gives mathematics its substance. Regardless of the point of view, mathematical objects provide a structural framework for the physical world.

Geometry is the undisputed language of the universe. Nature speaks the language of mathematics which has given rise to the more qualifying reference of simple geometry to the term of sacred geometry.

Galileo famously *said, "Mathematics is the language with which God has written the Universe".*

THE DIVINE PROPORTION

At the heart of sacred geometry is the Golden Ratio, also known as the Golden Cut, the Golden Rectangle, The Divine Proportion, and other names like the Golden Mean. Like PI π **3.14**, the Golden Mean is a strange number that we seldom question and often take for granted. This magical, infinite number **1.618** is represented by the Greek letter PHI $\boldsymbol{\phi}$. Different from PI, the golden mean goes unnoticed in our everyday life in things like buildings, plants, and living creatures, yet we find these things strangely pleasing to the eye.

In philosophy, especially that of Aristotle, the golden mean is the desirable middle between two extremes, one of excess and the other of deficiency. As an example, courage, a virtue, if taken to excess would manifest as recklessness, and if found to be deficient as cowardice. In Chinese philosophy, a similar concept called the Doctrine of the Mean was propounded by Confucius. Buddhist philosophy also includes the concept of the middle way.

To the Greek mentality this was considered to be an attribute of beauty. Both ancients and moderns realized a close association in mathematics between beauty and truth. The poet John Keats, in his *Ode on a Grecian Urn*, put it this way: "Beauty is truth, truth beauty, -- that is all ye know on earth, and all ye need to know."

The Greeks believed there were three ingredients to beauty: symmetry, proportion, and harmony. This triad of principles infused their life. They were attuned to beauty as an object of love and something to be imitated and reproduced in their lives, architecture, education, and politics, and they judged life by this mentality.

In mathematics, two quantities are in the golden ratio if their ratio is the same as the ratio of their sum to the larger of the two quantities.

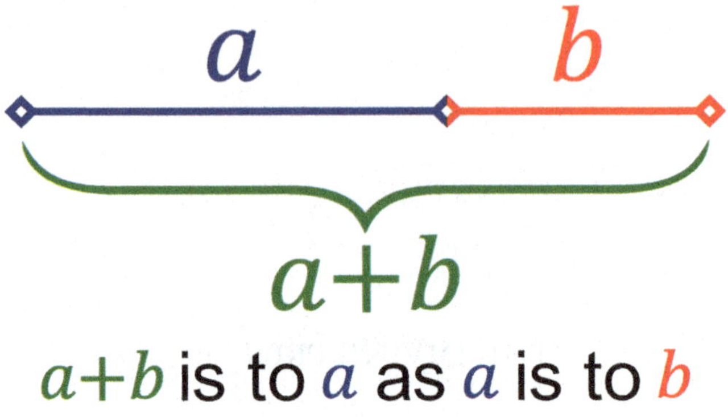

Line segments in the golden ratio

To construct a geometric representation of the golden ratio, divide a unit square into two equal rectangles with sides 1 and 1/2. Add the length of the diagonal in one rectangle to its short side. The new length is the golden ratio and the added on rectangle along with it plus the square are both "golden".

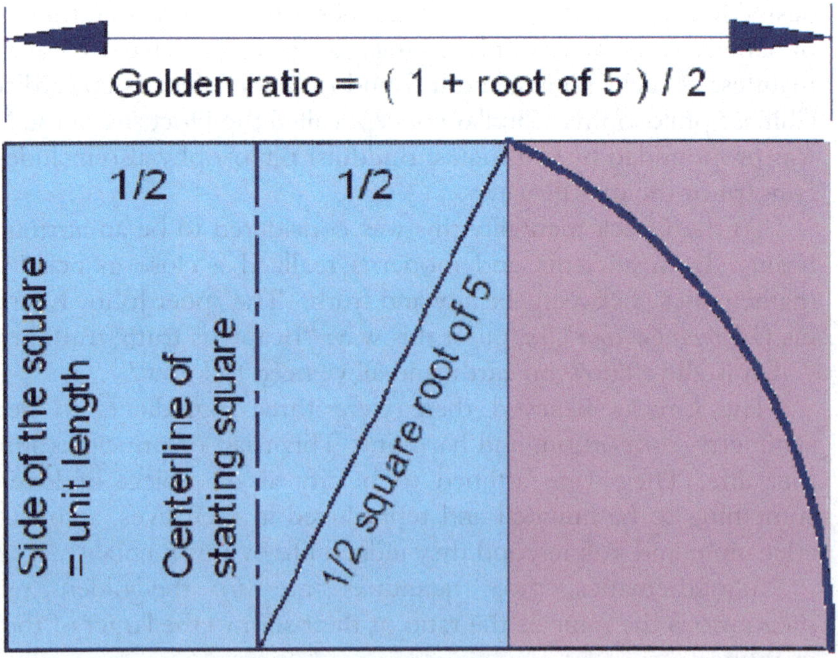

When you add a square along the long side of a golden rectangle, the new rectangle is again golden, and the next one, and so on expanding infinitely. If you draw a quarter circle into each successive square so that each arc begins where the prior one ends, you obtain a "logarithmic" spiral of self- replicating growth.

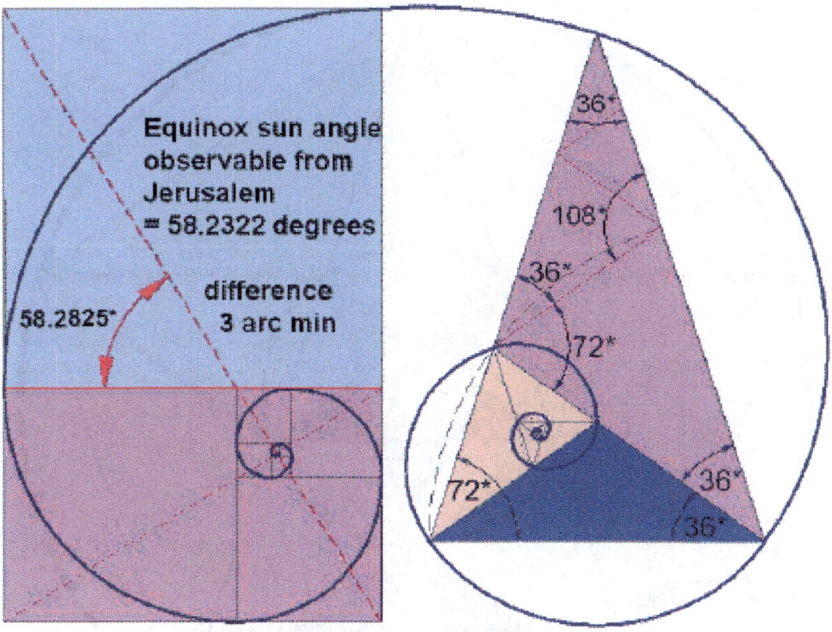

The golden ratio also yields two golden triangles, with side lengths 1, phi, and phi for the slender one, and 1, 1, and phi for the obtuse one. These fit alongside each other to reproduce their identical shapes again. When you connect the pointed ends of each successive obtuse triangle with an arc around its apex, you also create a logarithmic spiral.

Adding two obtuse golden triangles with their broad base along the equal sides of the slender one produces the pentagon, a rich source of golden ratios. Its diagonals are phi times its side length and divide each other in the same phi proportion to form a new pentagon.

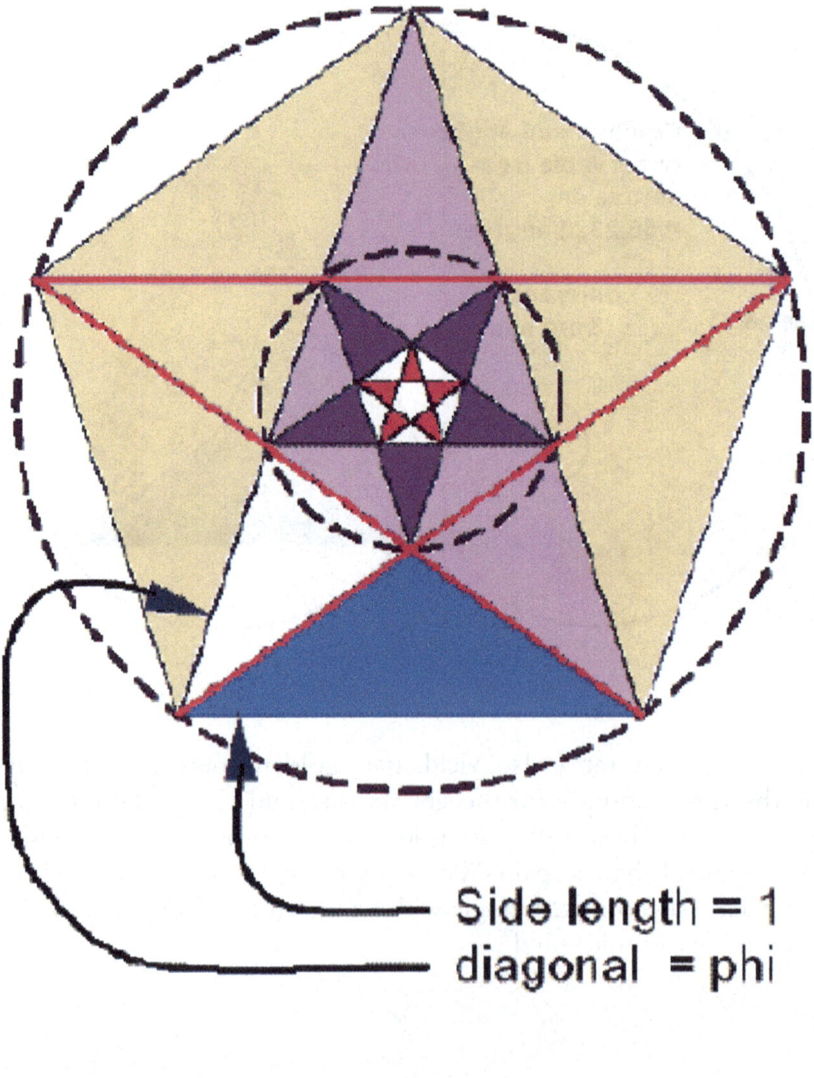

The Golden Mean models a precise mathematical geometric definition of a hologram that is made apparent through its ongoing expression in a sequence known as the Fibonacci spiral that encompasses everything that a spiral implies, including inspiration and respiration.

At the core of these expressions is the word spiration which is defined as the action of breathing as a creative or life-giving function of the Deity, or the act or manner that the Holy Spirit proceeds from the Father, or the Father and the Son, and the relation subsisting by virtue of this procession.

The diagram below is made up of squares, but the overall image is a rectangle that has the ratio of **1.618**. The curved lines within each of the squares are quarter circles, but as a whole they resemble the cross section of a sea shell and is in fact the same as the growth rate of the Nautilus Sea Shell - i.e. **1.618**.

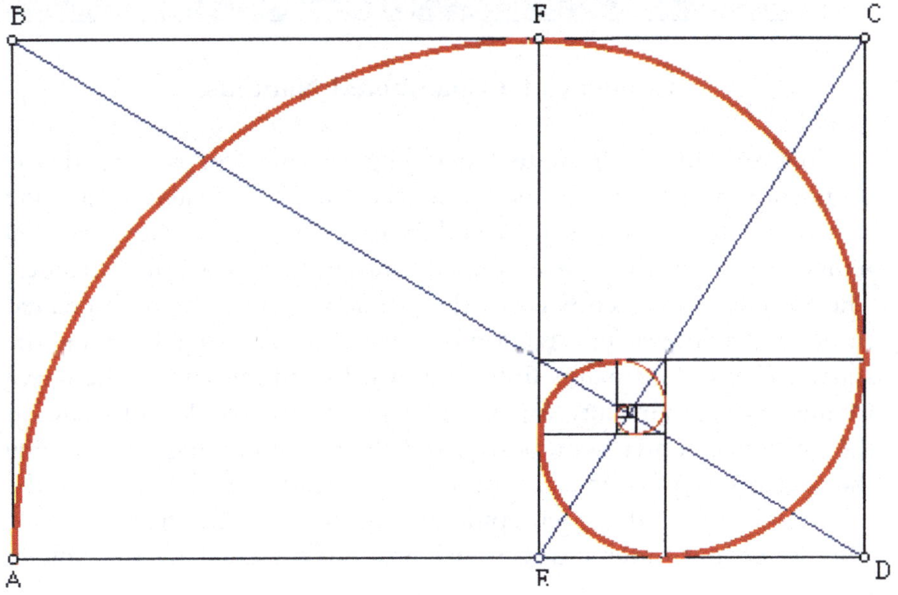

The Golden Mean Expressed as a Fibonacci Spiral

Cutaway of a Chambered Nautilus

Inherent in the infinite expanding progression of the divine proportion propagating in the Fibonacci Spiral is the square providing structure as the archetypal male and the curve or circle as the archetypal female. In nature this is exemplified through hurricanes or tornadoes. The ensuing vortex is inherently the geometry of how matter is created through a geometric contraction process of what is called curving the square, where the squares unfold outward to infinity while the curve rotates inward to infinity. Squares provide the matrix like substrate, or the primordial universal tapestry, and the act of curving and twisting the squares is the feminine process of creation which is why the Fibonacci ratio and golden spiral are intrinsic in all natural systems. This inner and outer bidirectional process is guided by the golden mean.

The expansion of the universe can be thought of as a masculine process of dispersion in a hyper-dimensional fractal cube-like fashion, and the formation of matter through a process of contraction, densification, and cohesion can be thought of as feminine. The very act of creation is fundamentally a feminine transmutation process of primordial substance into constructs.

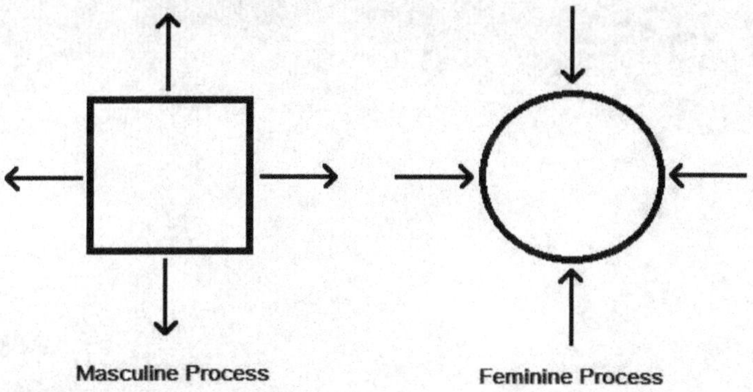

This spiral of creation is the core of inspiration which the Merriam-Webster dictionary defines as:

- A Divine influence or action on a person believed to qualify him or her to receive and communicate sacred revelation
- The action or power of moving the intellect or emotions
- The act of influencing or suggesting opinions
- The act of drawing in; *specifically*: the drawing of air into the lungs
- The quality or state of being inspired
- Something that is inspired: *A scheme that was pure inspiration*
- An inspiring agent or influence

A low pressure area shows a logarithmic spiral pattern.

The arms of spiral galaxies often have the shape of logarithmic spirals, like those shown here in the Whirlpool Galaxy.

Fibonacci discovered that if you start with the numbers **0** and **1** and add them together you get a new number -- in this case **1**. If you add the last number and the new number together, you get another new number, **2**. If you repeat this progression over and over again you end up with a long list of unique numbers known as the Fibonacci Series.

0, 1 — Added together gives a new number **1**.
0, 1, 1 -- Add the last two numbers and new number is **2**.
0, 1, 1, 2 -- Add the last two numbers and the new number is **3**.
0, 1, 1, 2, 3 -- Add the last two numbers and the new number is **5**.
0,1,1,2,3,5 -- The additions grow into a series of unique numbers.
0,1,1,2,3,5,8,13,21,34,55,89,144, 233,377 to infinity.

Starting from zero, if you take any two SEQUENTIAL numbers and calculate the ratio between them, an interesting pattern emerges.

1,0	Ratio	=	**1**	to	**0**	=	**0**
1,1	Ratio	=	**1**	to	**1**	=	**1**
2,1	Ratio	=	**2**	to	**1**	=	**2**
3,2	Ratio	=	**3**	to	**2**	=	**1.5**
5,3	Ratio	=	**5**	to	**3**	=	**1.6666**
8,5	Ratio	=	**8**	to	**5**	=	**1.6**
13,8	Ratio	=	**13**	to	**8**	=	**1.625**
21,13	Ratio	=	**21**	to	**13**	=	**1.61538**
34,21	Ratio	=	**34**	to	**21**	=	**1.61538**
55,34	Ratio	=	**55**	to	**34**	=	**1.61764**
89,55	Ratio	=	**89**	to	**55**	=	**1.6181**
144,89	Ratio	=	**144**	to	**89**	=	**1.6179**

If you continue in this manner the decimal figure will revolve around the magic number **1.618** because it is an infinite number.

From this self-apparent observation, the golden mean is a microcosm within the macrocosm that is expressed infinitely in both directions whether bigger or smaller in the same way that a hologram has the whole within all of its parts like an infinite hall of mirrors expressing reflection upon reflection.

Implicit within this overarching paradigm are the concepts of the implicate and explicate orders that describe two different frameworks for understanding the same phenomenon. The implicate, enfolded order is seen as a deeper, more fundamental order of reality. While the explicate or unfolded order includes the abstractions that humans normally perceive. The concept of implicate and explicate orders emphasizes the primacy of structure and process over individual objects.

Bohm and other physicists worked toward a model of quantum physics where the implicate order is represented in the form of an appropriate algebra or other pregeometry that considered spacetime as part of an explicate order connected to an implicate order they called *pre-space.* Others have suggested that an implicate order could be carried by an algebra, with the explicate order being contained in the various representations of this algebra.

This algebraic pregeometry is the golden mean.

Within this unfolding paradigm the brain is a hologram enfolded in a holographic universe, and all levels of creation are analogous to each other because each level of creation is an expression of the whole.

SACRED GEOMETRY

The term sacred geometry describes assertions of a mathematical order to the intrinsic nature of the universe which is supported by the fact that scientists see the same geometric, mathematical patterns arising from natural principles.

In sacred geometry, symbolic and sacred meanings are ascribed to certain geometric shapes and geometric proportions. In the ancient world certain numbers had symbolic meaning aside from their ordinary use for counting or calculating. Plane figures, polygons, triangles, squares, hexagons, and so forth were related to numbers in the same manner as the number three is to the triangle. Because they were visual, they carried more emotional value than the numbers themselves.

The belief that God created the universe according to a geometric plan has ancient origins. Plutarch attributed the belief to Plato, writing "Plato said God geometrizes continually".

The study of sacred geometry has its roots in the study of nature and the mathematical principles at work there like that seen in the chambered nautilus shell, low pressure weather patterns, and galaxies that all form logarithmic spirals.

The most prevalent traditional geometric forms ascribed to sacred geometry are:

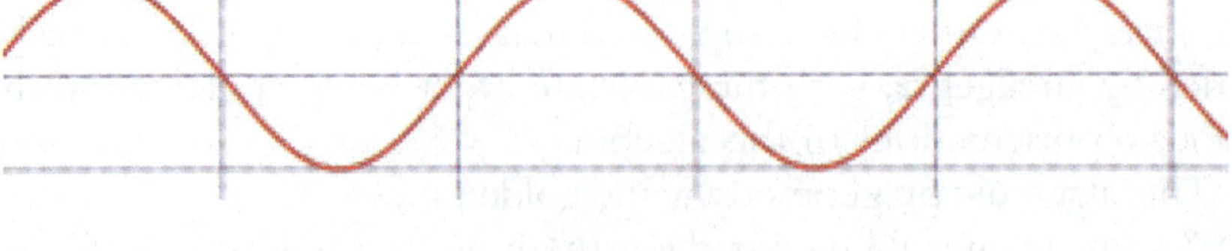

The Sine Wave

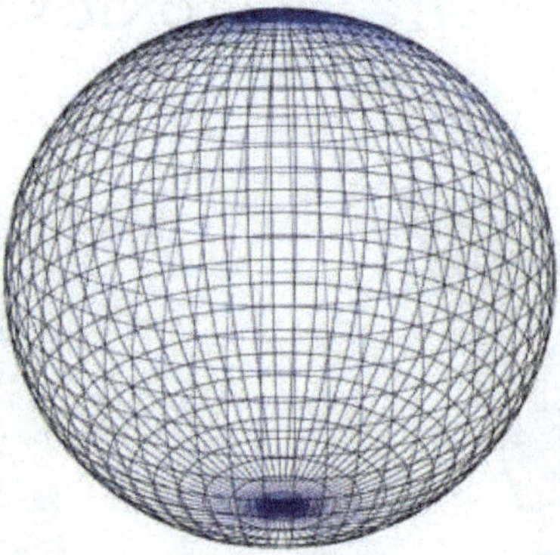

The Sphere

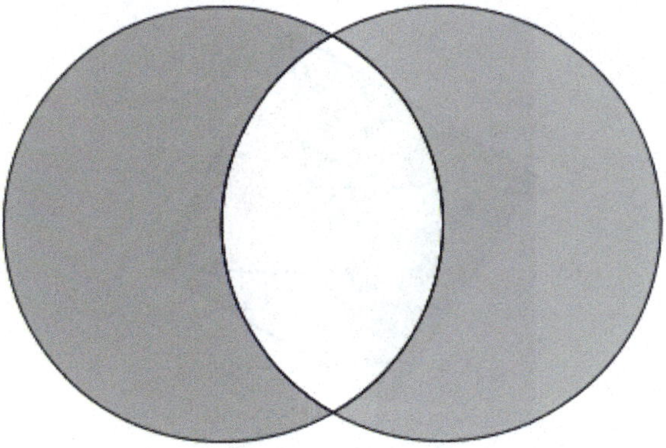

The Vesica Pisces

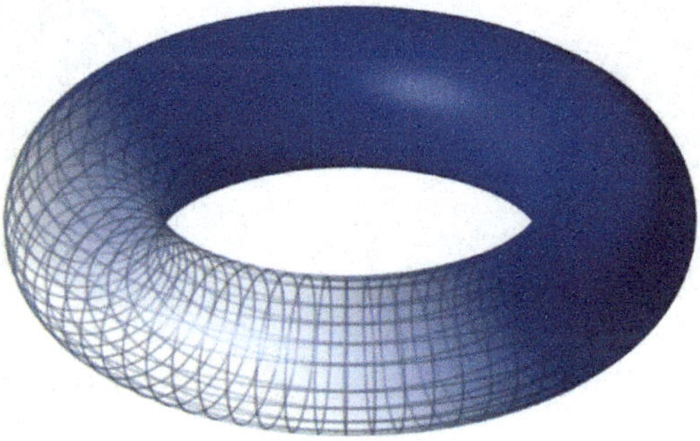

The Torus

THE 5 PLATONIC SOLIDS

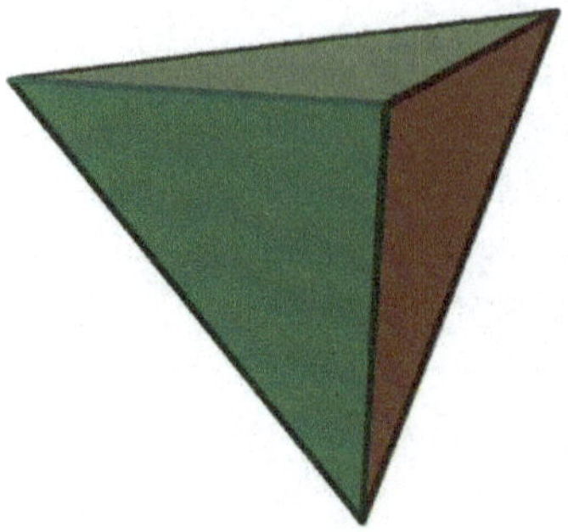

Tetrahedron (Four Faces)

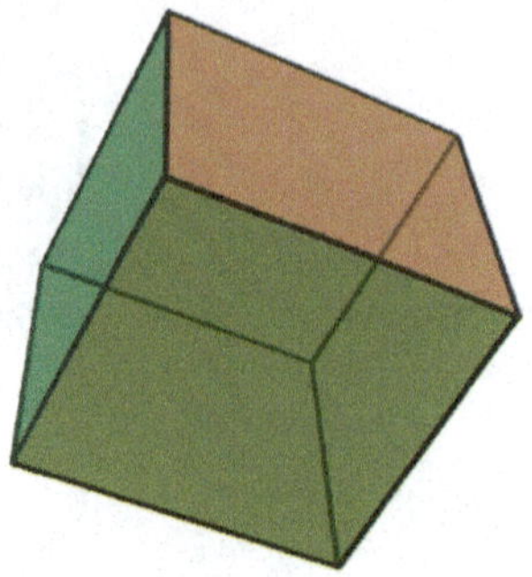

Cube or Hexahedron (Six Faces)

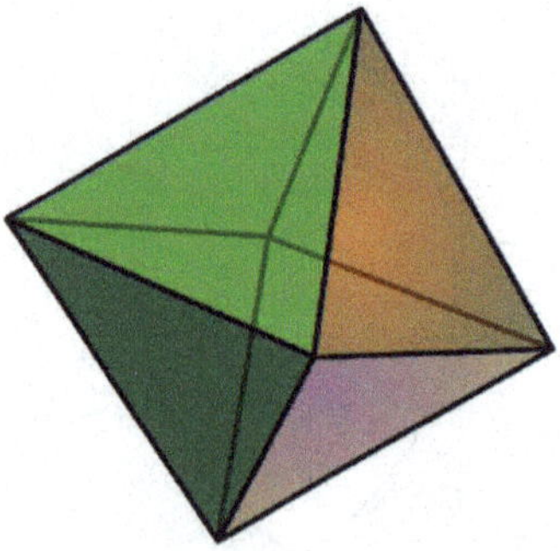

Octahedron (Eight Faces)

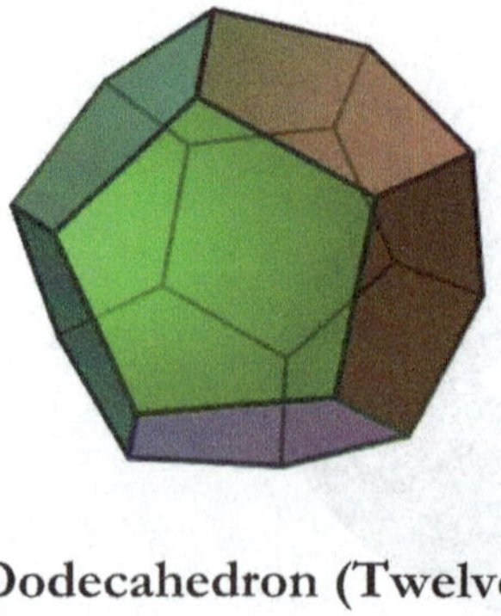

Dodecahedron (Twelve Faces)

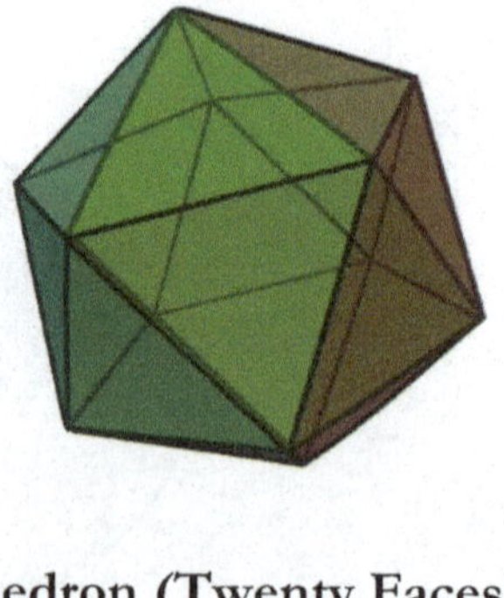

Icosahedron (Twenty Faces)

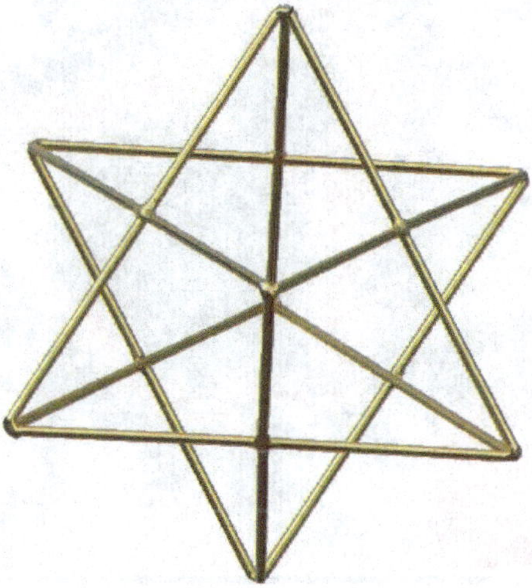

Star Tetrahedron

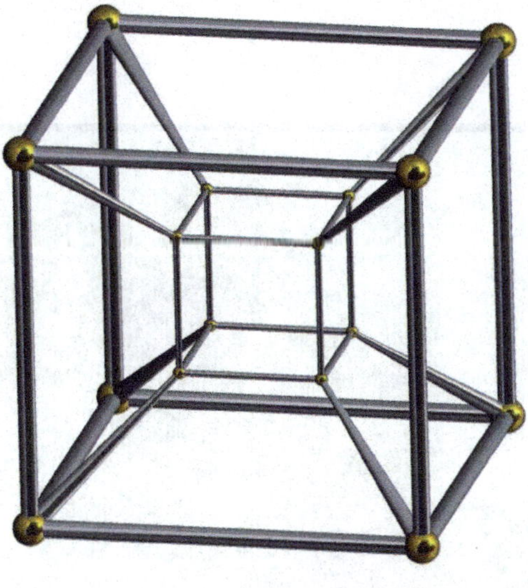

The Tesseract (4-dimensional cube)

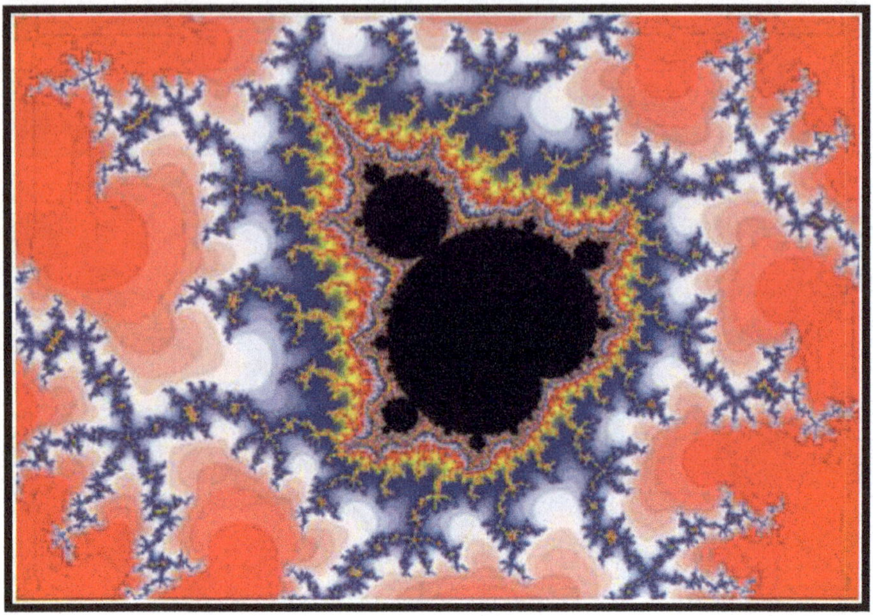

Fractals

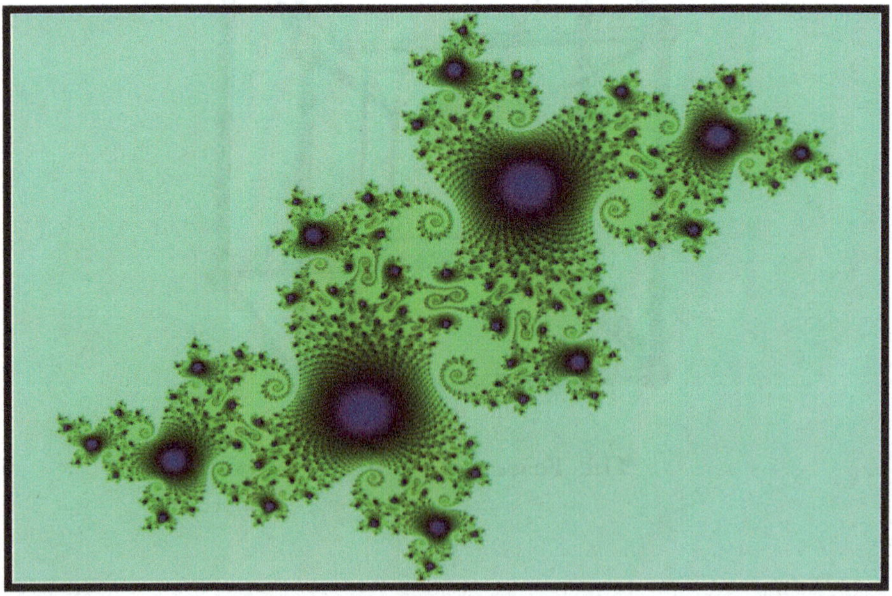

Aerial View of British Fractal Crop Circle

In man made expressions that reflect those found in the natural world, sacred geometry is used in the planning and construction of religious structures like churches, temples, mosques, religious monuments, altars, and tabernacles as well as sacred spaces like village greens, holy wells, and in the creation of religious art.

Cathédrale Notre-Dame de Chartres North Rose Window

Cathédrale Notre-Dame de Chartres South Rose Window

Cathédrale Notre-Dame de Chartres

Indian and Himalayan spiritual communities constructed temples and fortifications on design plans based on the mandala.

Mandala

Many of the sacred geometric principles of the human body and ancient architecture have been compiled into the well-known Vitruvian Man drawing by Leonardo Da Vinci, based on the much older writings of the Roman architect Vitruvius.

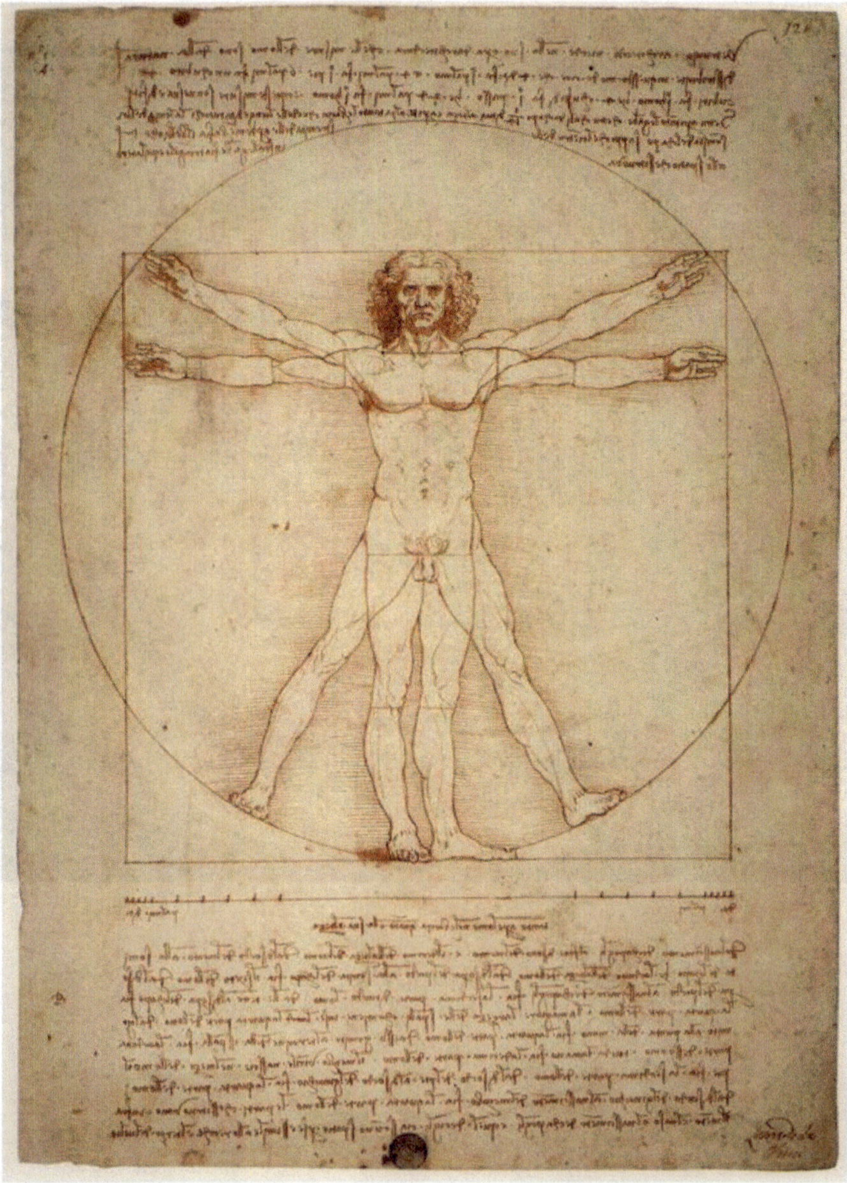

The Vitruvian Man

Ancient wisdom from many cultures tells us that our body is our temple. One of the most profound, densely packed storehouses of this wisdom still in existence on our planet is contained in the temple of Amun-Mut-Khonsu in Egypt, more commonly known as the Temple of Anthropocosmic Man.

The Temple of Anthropocosmic Man

This Temple of Anthropocosmic Man at Luxor is a masterpiece of art, science, and spirituality laid out in an elegant structure architecturally rendered to exhibit the proportions of Man within its design and artwork, as well as the mathematical and geometrical structure of the Cosmos and its locale within human consciousness. Pharaonic Consciousness not only recognized Man as the center of the Universe, it could formally equate it as well.

In his smaller work, *The Temple* in *Man*, that distills the massive two volume set *The Temple* of *Man*, R. A. Schwaller de Lubicz shares his observations of the temple, "…the canon proportions of the profile and the head, and of the head in relation to the body, are present. Here the Golden Number comes to our aid. It controls all vegetable and animal growth…"

He continues to write, "The outline of a human skeleton—traced according to anthropometrical methods and very carefully constructed, bone by bone—was superimposed on the general plan of the temple. The head (full face for the skeleton) is located exactly in the sanctuaries of the covered temple; the sanctuary of the barque of Amun is in the oral cavity; the clavicles are marked by walls; the chest is located in the first hypostyle of the covered temple and ends with the temple's platform. The abdomen is represented by the peristyle court, and the pubis is located exactly at the door separating this peristyle from the colonnade of Amun.

This marvelous colonnade is, in fact, dedicated to the femurs, the thighs; the knees are at the site of the gate in front of which sit the two colossi, marking the entrance to this colonnade. The tibias are in the court of Ramses, framed by the colossi, whose legs (tibias) are particularly pronounced. The little toe of our skeleton falls exactly at the northwest angle of the pylon. One might be tempted to think this skeleton had been constructed to be superimposed on the temple. But any skeleton, as long as it is harmonious (like the one represented here) can be projected thus on the plan of the temple and will coincide with it. Moreover, all the proportions of the skeleton may be checked against the actual measurements of the temple.

For my report, it was necessary to have recourse to the Egyptian canon; I have in this regard devoted a chapter to a subject that has never been dealt with until now—the importance the Ancients accorded the crown of the skull (cava).

This crown of the skull, marked off in Egyptian figuration by a headband, a diadem, a headdress or crown—is a veritable revelation with regard to psycho-spiritual knowledge of the Ancients. This is made clear by the placement of the principal organs of the Intellect and of all the control mechanisms of life in the various sanctuaries, whose figurations, texts, and architecture specify their purpose.

The Temple of Luxor is indisputably devoted to the Human Microcosm. This consecration is not merely a simple attribution: the entire temple becomes a book explaining the secret functions of the organs and nerve centers."

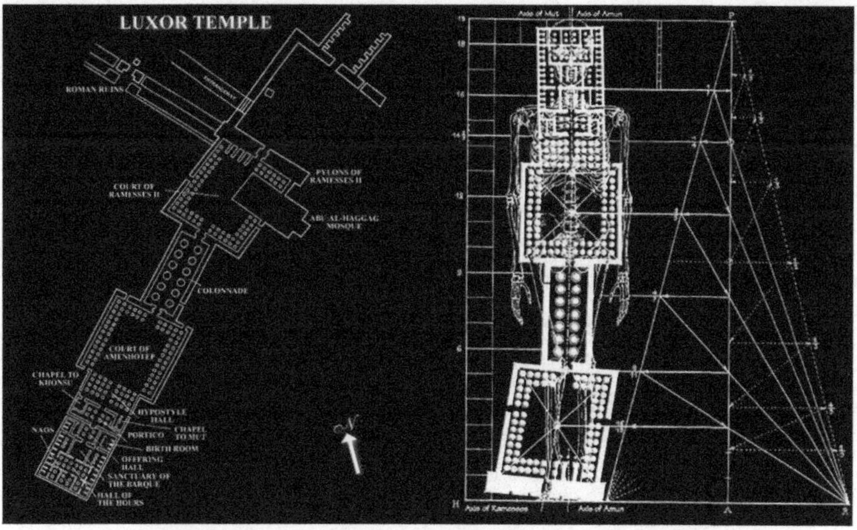

Mathematical Proportions of the Luxor Temple

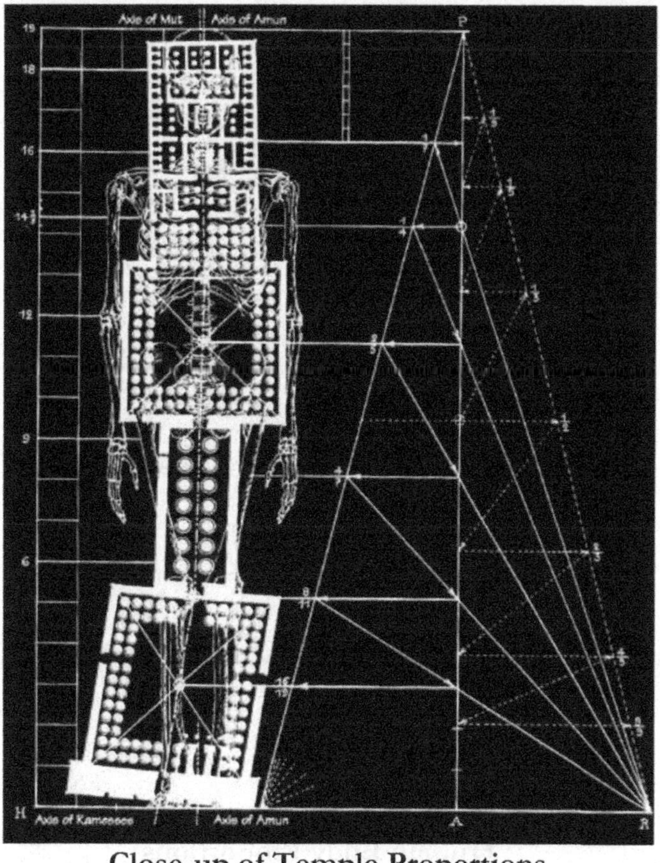

Close-up of Temple Proportions

The human body is phi-designed as the golden section template seen throughout our whole human form ratios, proving that humans, like the macrocosm of the planets and stars, and the microcosm of atomic and subatomic particles were all created using the PHI design.

The first example of the golden ratio in the average human body is apparent when the distance between the navel and the foot is taken as **1** unit, the height of a human being is equivalent to **1.618**.

Some other golden proportions in the average human body are:

- The distance between the fingertip and the elbow and the distance between the wrist and the elbow.
- The distance between the shoulder line and the top of the head and the head length.
- The distance between the navel and the top of the head and the distance between the shoulder line and the top of the head.
- The distance between the navel and knee and the distance between the knee and the end of the foot.

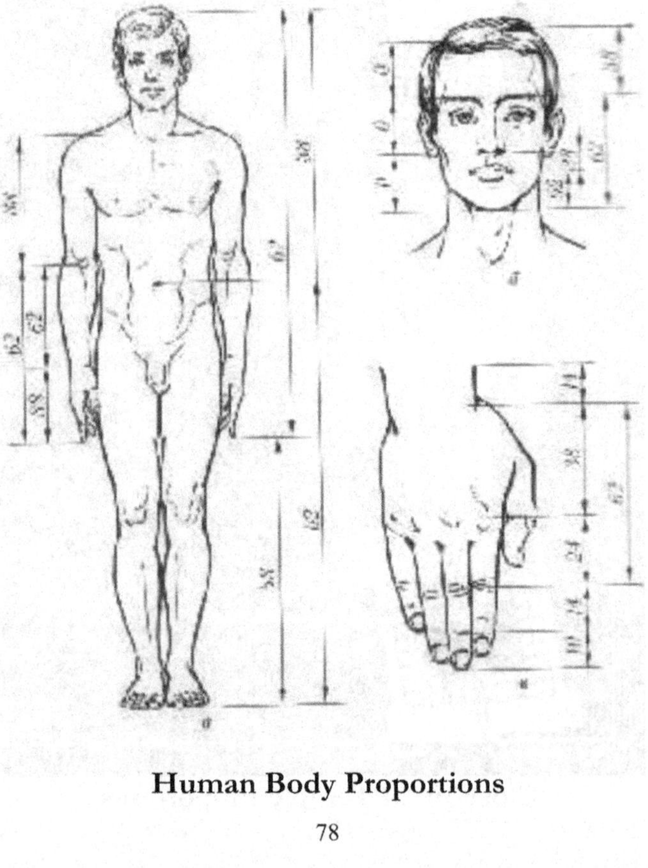

Human Body Proportions

Our fingers have three sections. The proportion of the first two to the full length of the finger gives the golden ratio, with the exception of the thumbs. The proportion of the middle finger to the little finger is also a golden ratio.

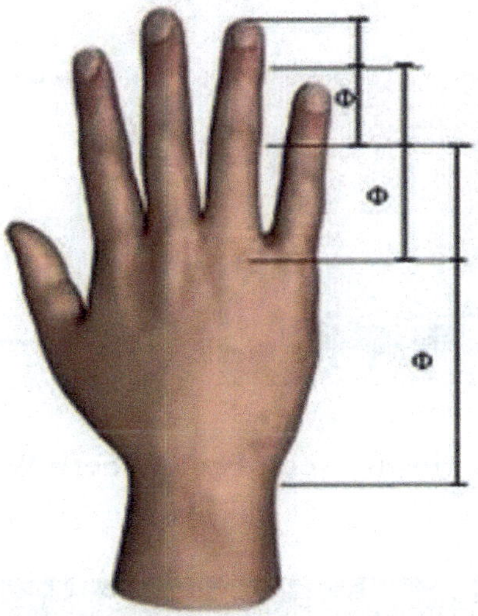

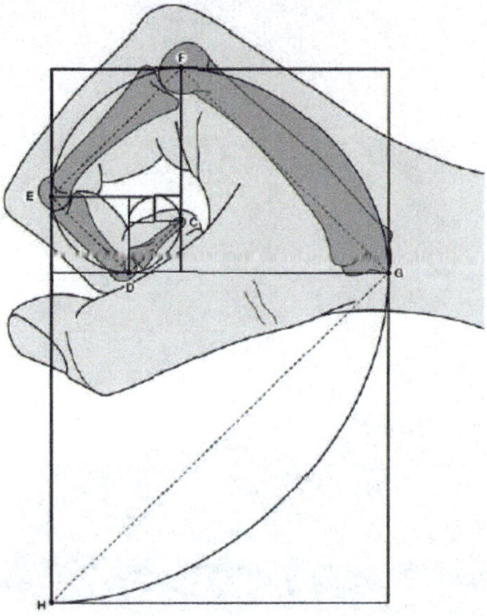

We have two hands, and the fingers on them consist of three sections. There are five fingers on each hand. Only eight of these are articulated according to the golden number: **2, 3, 5,** and **8** fit the Fibonacci numbers.

Likewise, there are several golden ratios in the human face. The total width of the two front teeth in the upper jaw over their height gives a golden ratio. The width of the first tooth from the center to the second tooth also yields the golden ratio of **1.618**.

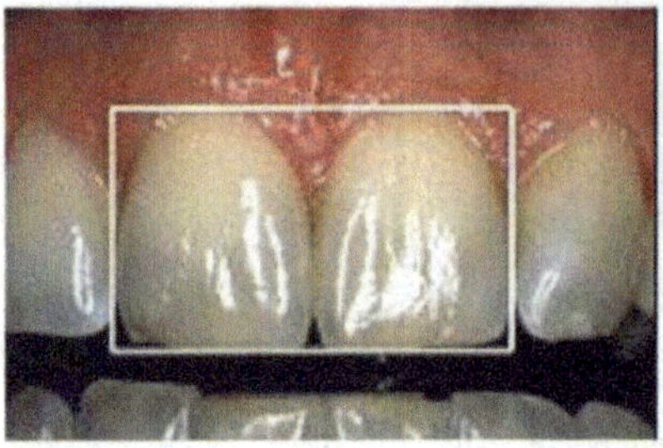

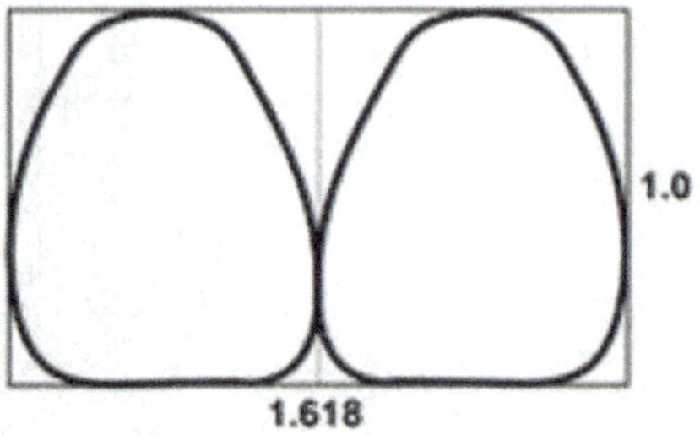

Golden Ratio Between Front Teeth Width and Height

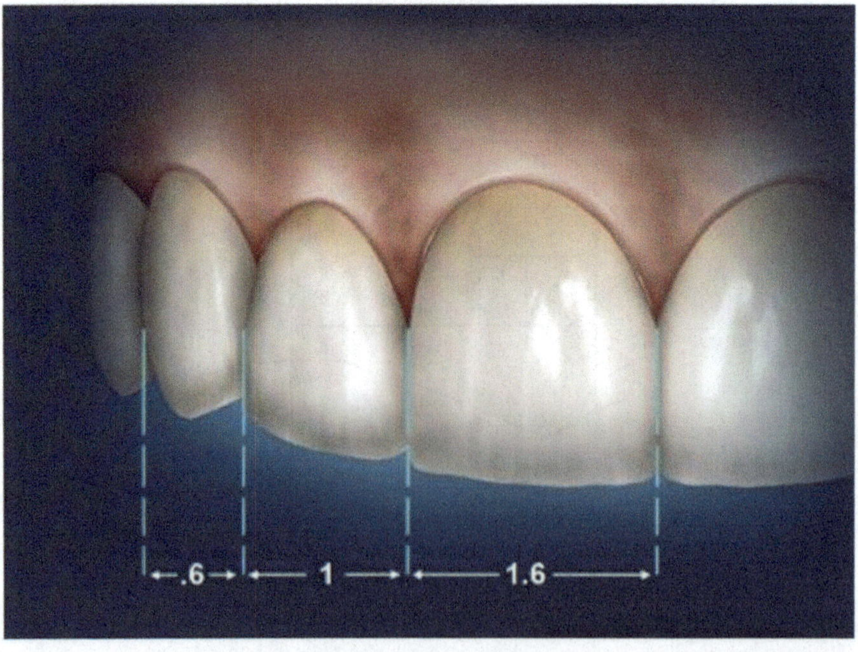

Golden Ration Between First and Second Tooth

Some other golden ratios in the human face are:

- Length of the face to the width of the face.
- The distance between the lips and where the eyebrows meet and the length of the nose.
- The length of the face and the distance between the tip of jaw and where the eyebrows meet.
- The length of the mouth and the width of the nose.
- The width of the nose and the distance between the nostrils.
- The distance between the pupils and the distance between the eyebrows.

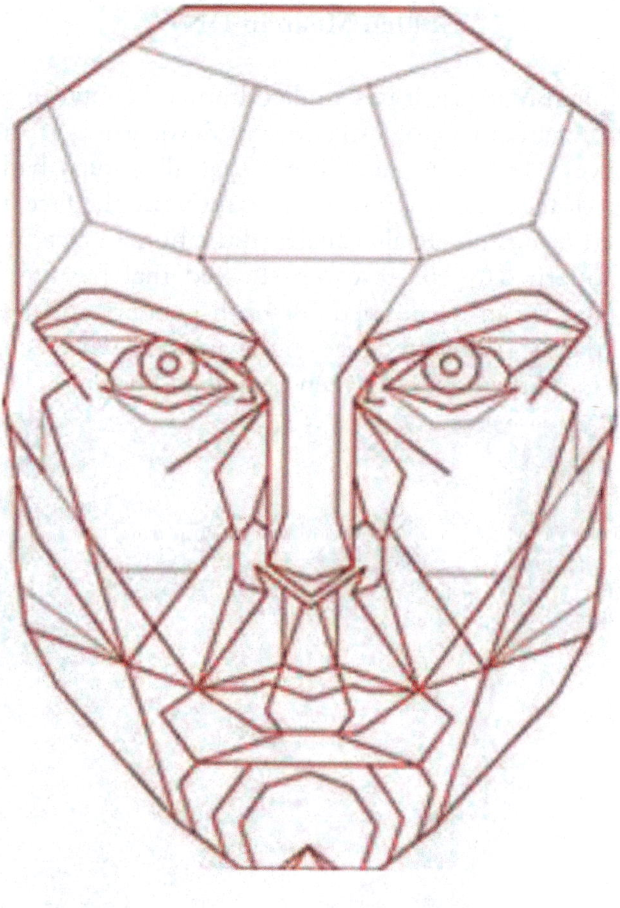

Golden Ratios in the Human Face

On a molecular level, the DNA molecule, which is the program for all life, is based on the golden mean which measures **34** angstroms long by **21** angstroms wide for each full cycle of its double helix spiral. **34** and **21** are numbers in the Fibonacci series and their ratio, **1.6190476** closely approximates phi, **1.6180339**.

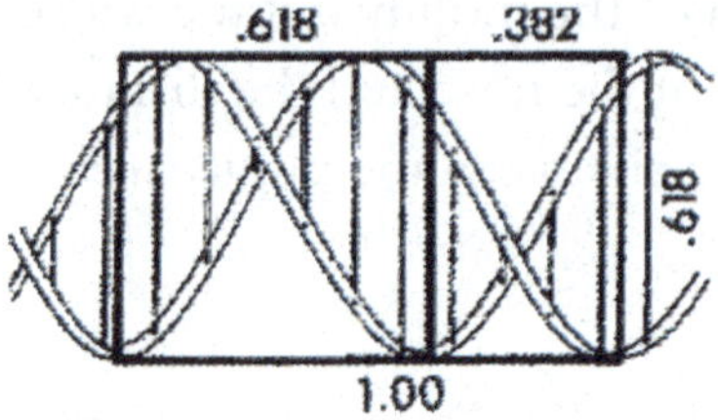

Golden Mean in DNA

Dr. Stephen Marquardt has studied human beauty for years in his practice of oral and maxillofacial surgery. Marquardt performed cross-cultural surveys on beauty and found that all groups had the same perceptions of facial beauty. The more symmetric the face, the more it is considered beautiful. He also analyzed the human face from ancient times to modern day. His research showed that beauty is not only related to phi, it can be defined for both genders and for all races, cultures, and eras, giving credence to the ancients who recognized the association in mathematics between beauty and truth.

Mathematical Association Between Beauty and Truth

Every physical attribute of a human being is geometrically definable in three dimensions according to the cosmic laws of sacred geometry and the divine proportion and these cosmic laws are self-apparent in geometric expression. Their truth is evident simply by examining them, giving credence to Keats' maxim, "Beauty is truth, truth beauty".

This sentiment has its roots in prehistory and can be seen in the enigmatic hieroglyphics and architecture of the ancient Egyptians, who saw no differences in science, art, and math. For them, these diverse disciplines were all one and the same.

This holographic paradigm is also inherent in the art of acupuncture which is based on the idea that every organ and bone in the body is connected to specific points in the body's surface. By activating these acupuncture points with needles or some other form of stimulation, diseases and imbalances affecting the parts of the body connected to them can be alleviated and cured. The human body has over a thousand acupuncture points organized in lines called meridians on the body's surface.

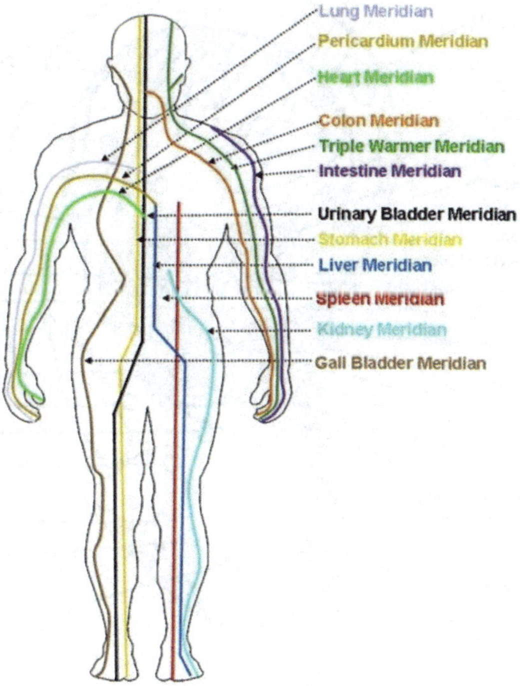

The Twelve Primary Meridians

In addition to the major acupuncture system, there are two smaller acupuncture systems on both ears called acupuncture microsystems that form an anatomical map of a miniature human inverted like a fetus referred to as the "little man in the ear". Dr. Terry Oleson, a psychobiologist at the pain Management clinic at the U.C.L.A. School of Medicine discovered that the ear microsystem can be used to accurately diagnose what is going on in the body. Increased electrical activity in one of the acupuncture points in the ear can indicate a pathological condition in the corresponding area of the body and it can reveal problems with the bones and internal organs. Just as every portion of a hologram contains the image of the whole, every portion of the body can also contain the image of the whole.

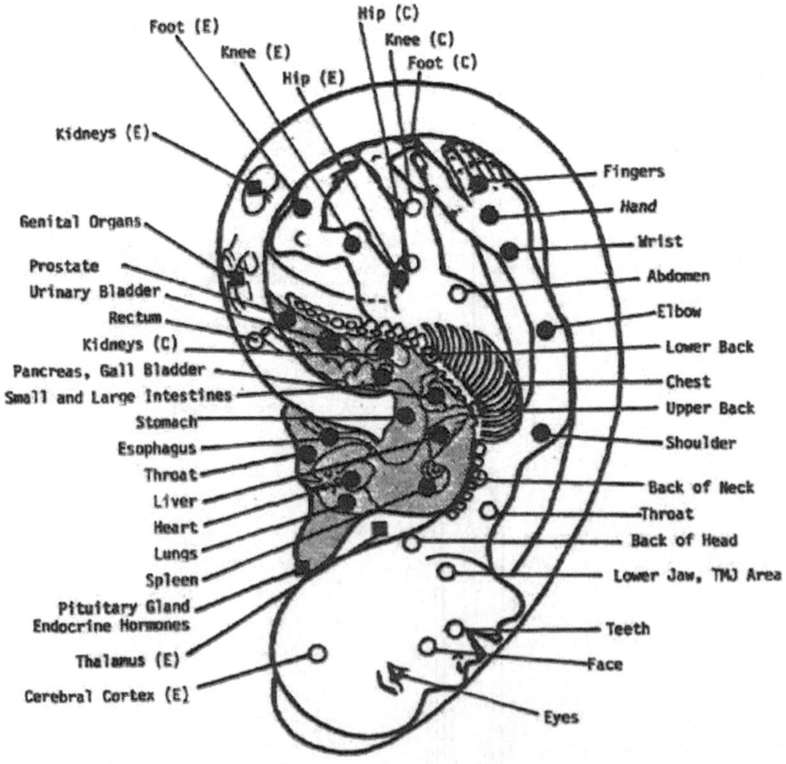

The Little Man In The Ear

Oleson says, "The ear holograph is, logically, connected to the brain holograph which is itself connected to the whole body, The way we use the ear to affect the rest of the body is by working through the brain holograph."

While director of the acupuncture education center in North Miami Beach, Florida, Dr. Ralph Allen Dale spent two decades tracking down clinical and research data from China, Japan, and Germany, accumulating evidence of 18 different microacupuncture holograms in the body, including ones in the hands, feet, arms, neck, tongue, and even the gums. Dale thought of these microsystems as holographic reiterations of the gross anatomy and hypothesized that every finger, and every cell contained its own acupuncture microsystem.

FROM PHYSICS TO METAPHYSICS

In a notion that parallels Bohm's implicate and explicate orders, the Sufis believed that the afterlife realm is the generative matrix that gives birth to the physical universe, and all things in physical reality arise from this spiritual reality. In this ancient reference to the non-local holographic qualities of reality, each of us contains the whole of heaven, and each of us contains the location of heaven, or as the Sufis put it, instead of having to search for spiritual reality "in the where," The "where" is *in* us. In discussing the nonlocal aspects of the afterlife realm, the Sufis called it the land of nowhere, the same sentiment expressed in the statement "the kingdom of heaven is within." The 12th century Sufis summed it up by saying, "the macrocosm is the microcosm," an earlier version of Blake's notion of seeing the world in a grain of sand.

Born in 1688, Emanuel Swedenborg was considered the Leonardo da Vinci of his era who studied science and was the leading mathematician in Sweden. Swedenborg spoke nine languages, was an engraver, a politician, an astronomer, and a businessman who built watches and microscopes as a hobby. He also wrote books on metallurgy, color theory, commerce, economics, physics, chemistry, mining, and anatomy, and invented prototypes for the airplane and the submarine.

Dr. George F. Dole, a professor of theology at the Swedenborg School of Religion in Newton, Massachusetts noted that one of the basic tenets of Swedenborg's thinking was that our universe is constantly created and sustained by two wave-like flows, one from heaven and one coming from our own soul or spirit. "If we put these images together, the resemblance to the hologram is striking," Dole said. "The intersection of two flows - one direct, from the divine, and one indirect, from the divine via our environment. We can view ourselves as interference patterns, because the inflow is a wave phenomenon, and we are where the waves meet."

The Hindus called the implicate level of reality Brahman which is formless, and the birthplace of all forms in visible reality that appear out of it and enfold back into it in endless flux. Like Bohm, who said that the implicate order can just as easily be called spirit, the Hindus personify this level of reality and say that it is composed of pure consciousness which is not only a subtler form of matter, but is more fundamental than matter. In the Hindu cosmogony it is matter that has emerged from consciousness, not the other way around, or as the Vedas put it, the physical world is brought into being through both the veiling and projecting powers of consciousness.

The Hindu Avatamsaka Sutra likens the universe to a legendary network of pearls said to hang over the palace of the God Indra and, "so arranged that if you looked at one pearl, you see all the others reflected in it." It further goes on to say, "in the same way, each object in the world is not merely itself, but involves every other object and in fact is everything else."

Like Bohm, who said that consciousness always has its source and the implicate, the aborigines believe that the true source of the mind is in the transcendent reality of dream time. The Dogon people of the Sudan also believe that the physical world is the product of a deeper and more fundamental level of reality and is perpetually flowing out of and streaming back into this more primary aspect of existence. As one Dogon elder described it, "to draw up and then return what one had drawn- that is the life of the world."

The Greek philosophers Anaximenes of Melitas, Pythagoras, Heraclitus, and Plato; ancient Gnostics; the pre-Christian Jewish philosopher Philo Judaeus; and the medieval philosopher Maimonides all embraced the macrocosm/microcosm idea.

After a shamanic vision of the subtler levels of reality the semi-mythical Egyptian prophet Hermes Trismegistus said that one of the main keys to knowledge was the understanding that "the without is like the within of things; the smallest is like the large." The medieval alchemists, for whom Hermes Trismegistus became a kind of patron saint distilled the sentiment into the motto "As above, so below." In talking about the same macrocosm equals microcosm idea, Hindu Visvasara Tantra states, "What is here is elsewhere."

The Oglala Sioux medicine man Black Elk put an even more non-local twist on the same concept. Standing on Harney Peak in the Black hills he witnessed a "great vision" during which he "saw more than I can tell and I understood more than I saw; for I was seeing in a sacred manner the shapes of all things in the spirit, the shape of all shapes as they must live together as one being." One of the most profound understandings he came away with after this encounter with the ineffable was that Harney Peak was the center of the world, but this distinction was not limited to Harney Peak. As Black Elk put it, "Anywhere is the center of the world." Over twenty five centuries earlier the Greek philosopher Empedicles wrote that, "God is a circle whose center is everywhere, and its circumference nowhere."

Fa-Tsang, the seventh century founder of the Hua-Yen school of Buddhist thought, when trying to communicate the ultimate interconnectedness and interpenetration of all things held that the whole cosmos was implicit in each of its parts. He also believed that every point in the cosmos was its center and likened the universe to a multi-dimensional network of jewels, each one reflecting all others infinitely. When the empress Wu did not understand what he meant by this image she asked him for further clarification. Fa-Tsang suspended a candle in the middle of a room full of mirrors. This, he told the empress, represented the relationship of the One to the many, then he took a polished crystal and placed it in the center of the room so that it reflected everything around it. This, he said, showed the relationship of the many to the One. Like Bohm who stressed that the universe is not simply a hologram, but a holomovement, Fa-Tsang stressed that his model was static and did not reflect the dynamism and constant movement of the cosmic interrelatedness among all things in the universe.

The 17th century German mathematician and philosopher Liebniz was familiar with the Hua-Yen school of Buddhist thought. Some have argued that this was why he proposed that the universe is constituted out of fundamental entities he called "monads", each of which contains a reflection of the whole universe. What is significant is that Liebniz also gave the world integral calculus, and it was integral calculus that enabled Dennis Gabor to invent the hologram.

Are we being dreamed by a single divine intelligence or are we being dreamed by the collective consciousness of all things? We cannot ask if the part is creating the whole, or the whole is creating the part because the part is the whole. Whether we call the collective consciousness of all things God or simply the consciousness of all things doesn't change anything. The universe is sustained by a tremendous act of ineffable creativity and it cannot be reduced to such limited simplistic terms.

Kalahari Bushmen say that the dream is dreaming itself.

THE NONLOCAL CENTER OF THE UNIVERSE

Ancient Egyptian Pharaonic Consciousness recognized Man as the center of the Universe and equated it within the proportions and geometry of the Temple of Anthropocosmic Man. This paradigm is borne out within our subjective experience of reality from an interference pattern that creates a holographic resonance inside of us within an "infinity zone" where consciousness springs into being through a unity of opposites that contain the implicate and explicate orders.

Like all things holographic, this phenomenon has its roots in the heart of sacred geometry through expressions of the infinite number **1.618** represented by PHI **ϕ**, the geometric representation of a hologram which also resides within the Vesica Pisces, referred to as The Mother OF All Form.

(Seat of Consciousness)

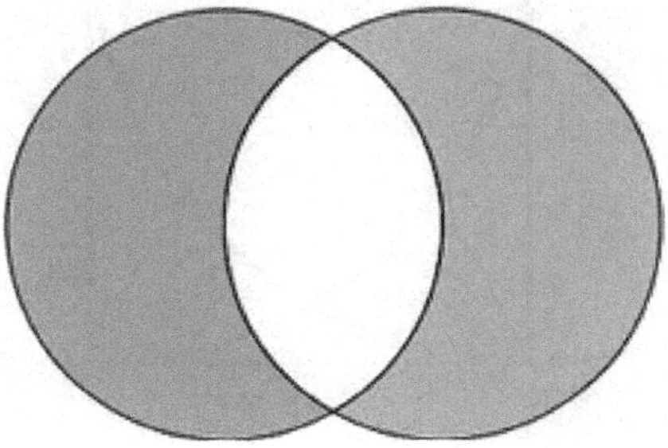

Subject (Inner Perception) **Object (Outer Perception)**

The Interference Pattern or Infinity Zone of the Vesica Pisces

Ancient scholars referred to this enigmatic figure of the Vesica Pisces as a mathematical shape formed by the intersection of two disks with the same radius, intersecting in such a way that the center of each disk lies on the perimeter of the other. In Latin, "vesica pisces" literally means "bladder of a fish", reflecting the shape's resemblance to the conjoined dual air bladders found in most fish.

When you add the mystical 3rd component of the 3rd circle, the Magic of the Trinity, you get phi, which is .618 known as Phi minus 1.

Phi = 1.618, and phi = .618.

This 3rd Circle, shown shaded, needed to have its radius touching the significant points of the Double Unit Square, whose diagonal is Root 5.

Root 2 is the diagonal of 1 Unit Square.

Root 5 is the diagonal of the Double Unit Square.

Root 3 is the vertical distance touching the 2 intersecting circles.

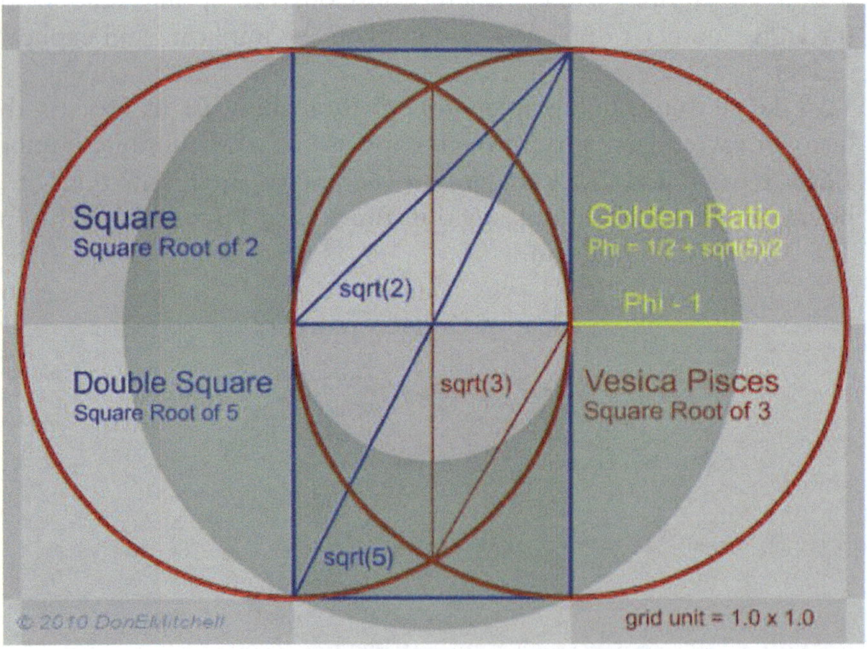

Phi Within the Vesica Pisces

Gurdjieff said that before examining the laws of transformation of Unity into Plurality we have to examine the fundamental law that creates all phenomena in all the diversity or unity of all universes, which he dubbed the "Law of Three", or the law of the *three principles* or *three forces.* Every phenomenon on whatever scale and in whatever world it takes place, from molecular to cosmic, is the result of the combination or meeting of three different and opposing forces. The first force is called active or positive, the second, passive or negative, and the third, neutralizing.

The creation of a permanent third principle for man is the transformation of duality into trinity. This can be seen in the permutation of duality and the basis of the expression of life when the circle that encompasses "all that is" moves in a primary scission that divides itself into the Vesica Pisces. In this dualistic split, 1 into 2 instantly becomes 3.

Our conscious awareness lives in this middle ground that represents the focal point and fulcrum between subjectivity and objectivity, or spirit and matter, and we have a front row seat where we control the interpretation and what we choose to manifest as a result of that stimuli. This is a zero point, ground zero, the eye of the storm, and the center of the universe, which is *nothing* - a place of power.

The Infinity Zone.

In the earliest traditions the supreme being was represented by a sphere, the symbol of a being with no beginning, no end, continually existing, perfectly formed, and profoundly symmetrical. The addition of a second sphere represented the expansion of unity into the duality of male and female, or god and goddess. By overlapping, the two spheres, the god and goddess created a divine offspring.

The Vesica Pisces motif and its derivatives, the Flower of Life, Tree of Life, and fundamentals of geometry have thousands of years of history that predates all major religions.

The son or daughter of the god and goddess is associated with the overlapping of the spheres. In the case of Jesus Christ, the two dimensional figure also served as a symbol for the miracle of the fishes. The "tail" also served to more easily identify the source of the plane figure. This also conveys the spiritual power originating from the interior of this symbol.

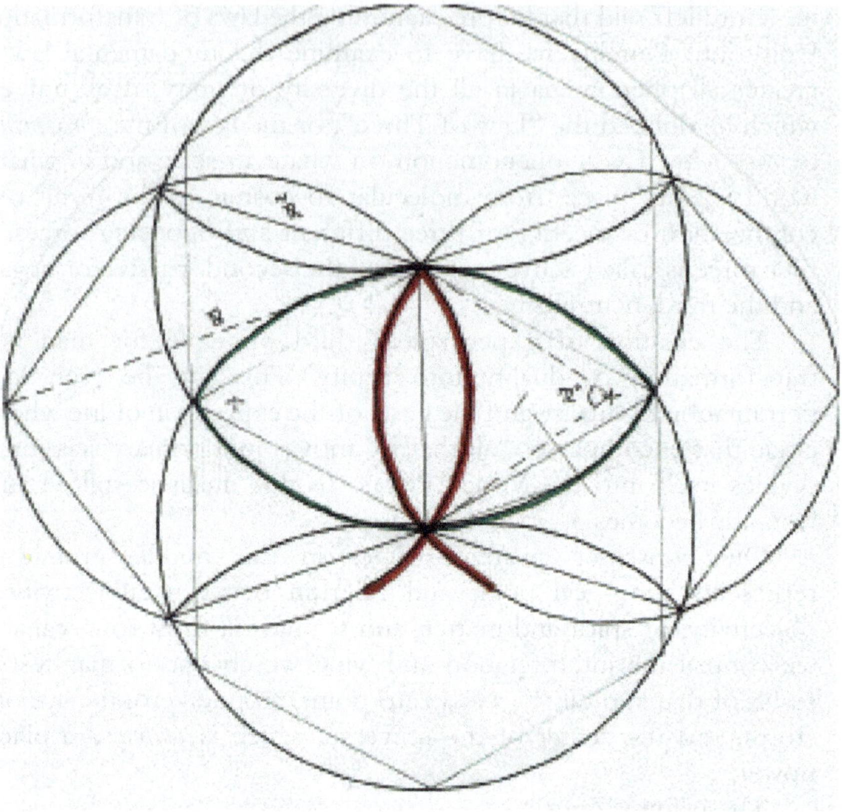

The intersection of two overlapping spheres, including their interior portion, along with the more common two dimensional version represents among other things:

- The joining of God and Goddess to create an offspring
- A symbol for Jesus Christ
- In art a pointed oval used as an aureole in medieval sculpture and painting
- The vagina of the female goddess
- The basic motif in the Flower of Life
- An overlay of the Tree of Life
- The formative power of polygons
- A geometrical description of square roots and harmonic proportions
- 7) A source of immense power and energy

Virtually every medieval church in Europe uses a standard motif of the Vesica Pisces in two dimensions. The fact that many of these churches were dedicated to the Virgin Mary or to Mary Magdalene, known as the goddess, is part of the understanding. Several of the churches in northern France are even located in such a manner that their *points of light* recreate the "lights" of the constellation Virgo. In Glastonbury, England, the site normally attributed to Avalon, the island of the Goddess, is where the Chapel of St. Mary is located, which is patterned with the use of the Vesica Pisces.

The goddess of any and all religions that recognize her power and significance use the Vesica Pisces to identify her. From the overlapping pools of water and the chalice well cover in the goddess's garden in Glastonbury to any number of representations of the Tree of Life, the goddess and her ability to create and birth life are celebrated.

Drunvalo Melchizedek, in his Flower of Life symbolism, uses the Vesica Pisces and considers it the geometric image through which light was created. One glance at the Flower of Life makes it apparent that this symbol contains the implicate and explicate orders within it, making it another perfect mathematical geometric representation of a hologram.

Flower of Life

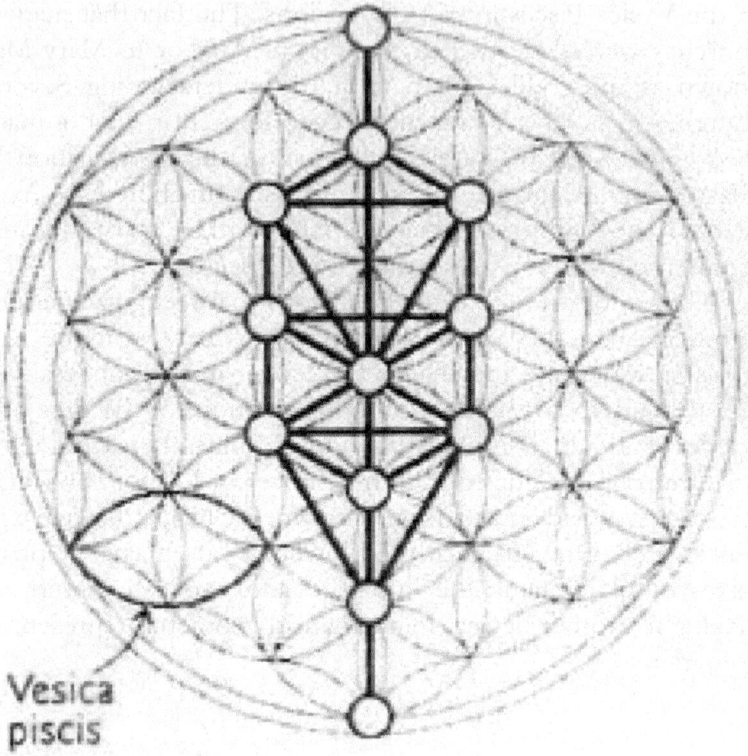

Tree of Life

The Tree of Life includes the Vesica Pisces and Flower of Life. The ease with which the patterns fit make the inclusion automatic.

Robert Lawlor, in *Sacred Geometry: Philosophy & Practice,* notes that the Ö3, also known as the ORF Eye contained within the Vesica Pisces is "the formative power giving rise to the polygonal 'world'."

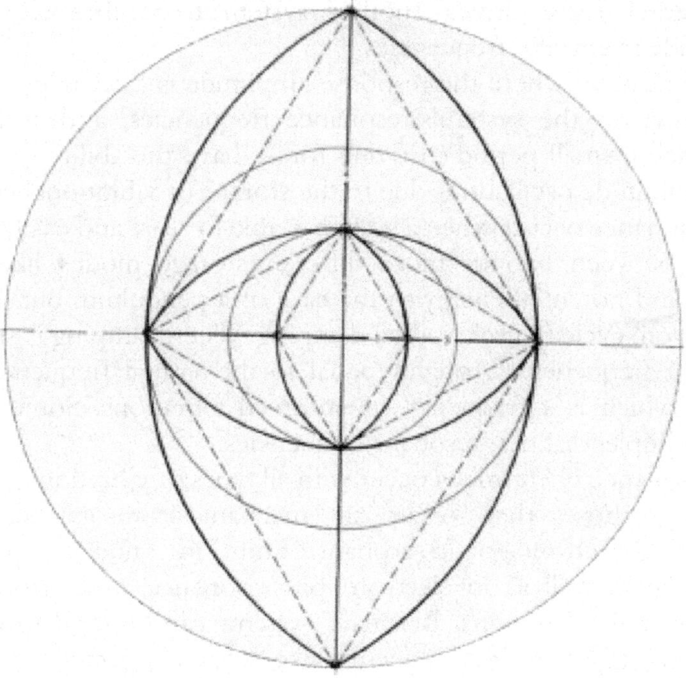

Additionally, the square roots of 2, 3, and 5, three of the first digits in the Fibonacci Numbers can be geometrically calculated. This is just an inkling of the possibilities and relationships that can be explored.

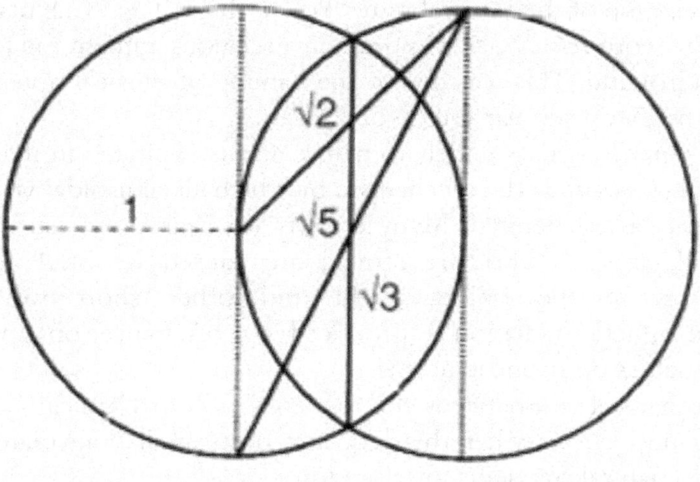

In physics, resonance is a phenomenon where a vibrating system or external force drives another system to oscillate with greater amplitude at specific frequencies.

Frequencies where the response amplitude is a relative maximum are known as the system's resonance frequencies, and at resonant frequencies, small periodic driving forces have the ability to produce large amplitude oscillations due to the storage of vibrational energy.

Resonance occurs when a system is able to store and easily transfer energy between two or more different storage modes like kinetic energy and potential energy in the case of a pendulum, but there are losses from cycle to cycle, called damping. When damping is small, the resonant frequency is roughly equal to the natural frequency of the system which is a frequency of unforced vibrations. Some systems have multiple, distinct, resonant frequencies.

Resonance phenomena occur with all types of vibrations or waves, including those that occur in mechanical resonance, acoustic resonance, electromagnetic resonance, and in nuclear magnetic resonance, as well as in electron spin resonance, and resonance of quantum wave functions. Resonant systems can be used to generate vibrations of a specific frequency like what happens with musical instruments, or it can pick out specific frequencies from a complex vibration containing many frequencies as in filters.

The term *resonance* comes from the Latin *resonantia*, "echo". From *resonare*, "resound" originates from the field of acoustics, and is observed in musical instruments, like when strings vibrate and produce sound without direct excitation by the player.

The name of the musical note "Re" in the solfège scale may come from the word *resonare*, as it appears in a religious anthem for John the Baptist in Latin. This was due to the naming of musical notes by the Italian medieval scholar Guido of Arezzo.

Resonance occurs widely in nature and is exploited in many man made devices and is the mechanism by which all sinusoidal waves and vibrations are generated. Many sounds we hear when hard objects of metal, glass, or wood are struck, are caused by brief resonant vibrations in the object. Light and other short wavelength electromagnetic radiation is produced by resonance on an atomic scale, such as electrons in atoms.

Mechanical resonance is the tendency of a mechanical system to absorb more energy when the frequency of its oscillations matches the system's natural frequency of vibration.

Acoustic resonance is a branch of mechanical resonance

concerned with the mechanical vibrations across the frequency range of human hearing. When sound waves strike a wall in an auditorium the sound reacts to various frequency energy dependent on the composition and various geometrical degrees present in the wall. Acoustic resonance is an important consideration for instrument builders, as most acoustic instruments use resonators like the strings and body of a violin, the length of tube in a flute, and the shape of and tension on a drum membrane.

Electrical resonance occurs within electric circuits at resonant frequencies when the impedance of the circuit is at a minimum in a series circuit or at maximum in a parallel circuit, or when the transfer function is at a maximum. Resonance in circuits are used for transmitting and receiving wireless communications in computer networks, television, cell phones, and radio.

When two or more objects are vibrating at the same frequency there is a maximum exchange of energy between them and they share the same characteristics of that particular energy made evident by their synchronization which is the coordination of events to operate a system in unison. This is what the conductor of an orchestra does to keep the orchestra *in time.* Systems operating with all their parts in synchrony are said to be *synchronous* or *in sync.*

Synchronization is an emergent property that occurs in a broad range of dynamical systems, including neural signaling, evident in the synchronized activity of neural ensembles that give rise to macroscopic oscillations of Alpha waves, the beating of the heart, and the synchronization of fire-fly light waves.

In groups, synchronization of movement has been shown to increase conformity, cooperation, and trust. In dyads which are groups of two people, synchronization increases affiliation, self-esteem, compassion, and altruistic behavior as well as increased rapport.

During arguments, synchrony between the arguing pair has been noted to decrease, but it is not clear whether this is due to the change in emotion or other factors. Evidence has shown that movement synchronization requires other people to cause its beneficial effects, as the effect on affiliation does not occur when one of the dyad is synchronizing their movements to something outside of the dyad. This is known as interpersonal synchrony.

Research in this area detailing the positive effects of synchrony have attributed this to synchrony alone, but many of the experiments incorporate a shared intention to achieve synchrony. The

Reinforcement of Cooperation Model suggests that perception of synchrony leads to reinforcement that cooperation is occurring, which leads to the pro-social effects of synchrony.

Synchronization occurs in a broad range of dynamical systems, including mechanical, acoustic, electrical, and neural signaling, and has been shown to increase conformity, cooperation, pro-social effects and trust in groups. We can also synchronize our inner universe so that it comes into resonance both internally and externally with the world and people around us, as well as in the plant and animal kingdoms along with other normally invisible energies that surround and inform us.

Discoveries in quantum physics lay the ground for this by proving that matter is more empty space than solidity. In terms of Einstein's $E=MC^2$, matter is another form of energy, and energy is the essence of consciousness. The fundamental nature of this paradox tells us that positive and negative are extremes. It is the energy between them that matters which is where we exist within ourselves; at the threshold of subject and object, in the center, of our duality. When we transcend the poles, we become centered in the Infinity Zone and rise above the opposition of forces to embrace the energy that they manifest.

When the mid-point rises above the baseline, its transcendence creates a triad, trinity, or triangle.

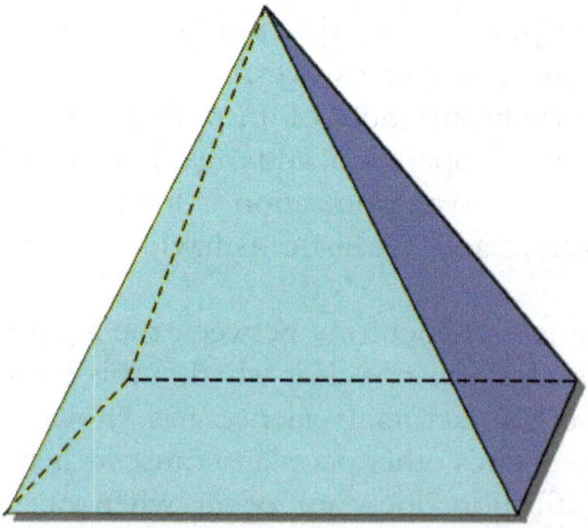

Pyramid Symbolizing Transcendence

The three dimensional representation of a triangle with four sides creates a pyramid, one of the greatest enigmas known to mankind. The square base of a pyramid represents the four dimensions and the four directions of the physical world that rise up pointing toward the heavens in a diminishing apex that represents an infinity point where matter ends and spirit begins.

If you consider the wave particle phenomenon and the fact that everything is energy, in conventional thinking this "back and forth" instantaneous transformation is seen as a transition from energy, or spirit, something without mass, into matter.

The interference or infinity pattern of subjective, subject-object perception is also thought of as the basis for telepathy, resonance, animal communication, and any method of tuning in that relates to the carrier frequency or source beam of a hologram in the same way a radio station transmits a signal that multiple receivers can tune into to hear the same content simultaneously.

The observed phenomenon of the appearance of a point or particle is energy manifesting into the physical realm. This coupled with the fact that it moves back and forth like the flickering image of a movie defines it as a wave-like pattern. The duality paradox of this separate, yet unified manifestation between the energetic and physical realms is an interference pattern that contains the implicate and explicate orders that define a hologram.

MORPHIC RESONANCE AND MORPHIC FIELDS

Biologist Rupert Sheldrake is best known for his hypothesis of morphic fields and morphic resonance that leads to a vision of a living developing universe with its own inherent memory. Over the course of fifteen years of research on plant development he came to the conclusion that for understanding the development of plants, their morphogenesis, genes, and gene products were not enough. Morphogenesis also depends on organizing fields.

How *does* a massive oak grow out of an acorn?

The same arguments apply to the development of animals. Since the 1920s many developmental biologists have proposed that biological organization depends on fields, variously called biological fields, developmental fields, positional fields, or morphogenetic fields.

A number of organisms live as free cells, including many yeasts, bacteria, and amoebas. Some form complex mineral skeletons, like in diatoms and radiolarians. Making the right proteins at the right times doesn't explain the complex skeletons of these structures without other forces coming into play, including the organizing activity of cell membranes and microtubules.

Most developmental biologists accept the need for a holistic or integrative conception of living organization, otherwise biology will continue floundering, even drowning in oceans of data, as more genomes are sequenced, genes are cloned, and proteins are characterized.

Sheldrake suggested that morphogenetic fields work by imposing patterns on otherwise random or indeterminate patterns of activity as is the case when they cause microtubules to crystallize in one part of the cell rather than another, even though the subunits that they are made from are present throughout the cell.

Morphogenetic fields are not fixed, they evolve. The fields of

Afghan hounds and poodles have become different from their common ancestors, wolves. How are these fields inherited? It's been postulated that they are transmitted from past members of the species through a kind of non-local resonance, called morphic resonance.

The fields organizing the activity of the nervous system are also inherited through morphic resonance, conveying a collective, instinctive memory. Each individual draws upon and contributes to the collective memory of the species, meaning new patterns of behavior can spread faster than would otherwise be possible. As an example, if rats of a particular breed learn a new trick in Harvard, then rats of that breed should be able to learn the same trick faster all over the world. There is evidence from laboratory experiments that this actually happens.

Social groups are also organized by fields, as in schools of fish and flocks of birds. Human societies have memories that are transmitted through the culture of the group and communicated through the ritual re-enactment of a founding story or myth, as in the Jewish Passover celebration, the Christian Holy Communion, and the American thanksgiving dinner, where the past become present through a resonance with those who have performed the same rituals before.

If we want to stick to the idea of natural laws, we can say that as nature itself evolves, the laws of nature also evolve similar to the way human laws evolve over time, but how would natural laws be remembered or enforced?

Habits are less human-centered. Many kinds of organisms have habits, but only humans have laws. The habits of nature depend on non-local similarity reinforcement. Through morphic resonance, the patterns of activity in self-organizing systems are influenced by similar patterns in the past, giving each species and each kind of self-organizing system a collective memory.

Habits are subject to natural selection and the more often they are repeated, the more probable they become, other things being equal. Animals inherit the successful habits of their species as instincts. We inherit bodily, emotional, mental and cultural habits, including the habits of our languages.

The morphic fields of social groups connect members of the group even when they are miles apart and provide channels of communication through which organisms can stay in touch at a distance. They also help provide an explanation for telepathy. There is good evidence that many species of animals are telepathic, and telepathy seems to be a normal means of animal communication that is normal, not paranormal, natural, not supernatural, and is common between people, especially those who know each other well.

The morphic fields of mental activity are not confined to the inside of our heads. They extend beyond our brain though intention and attention. We are already familiar with the idea of fields extending beyond the material objects that they are rooted in. Magnetic fields extend beyond the surfaces of magnets, the earth's gravitational field extends beyond the surface of the earth keeping the moon in its orbit, and the fields of cell phones stretch out beyond the phone itself. In this same manner, the fields of our minds extend far beyond our brains.

This "invisible" method of communication and transmission through resonating fields also extends beyond individual organisms, breeds, species, colonies, social groups, and other self-organizing systems to interact with each other.

In a shamanic world view that existed thousands of years before modern day "discoveries" of quantum physics, *everything* is seen as energy, and the word spirit and energy have the exact same meaning. If you look at reality from this perspective, how do plants, animals, insects, and other diverse species communicate with us and each other?

On an olfactory level, plants call out from a distance by their varied aromas that form the basis for scents and perfumes that are attractive to humans, often to the point of stimulating a resonance between lovers. Additionally they are used as powerful tools in aromatherapies to induce recall, reinforce powerful emotion, or simply for the calming quality of their fragrances.

When they affect us in these profound ways, it is because we resonate with them.

On a visual level we are attracted to them by their stunning beauty, and we display them in our homes and businesses to "brighten things up." It is a long held tradition to bring roses to a romantic partner to show love as a traditional courting strategy, and we send flowers to commemorate special human occasions like birth, marriage, and death.

These visual and olfactory cues combine to attract pollinators and other synergistic interactions to spread seeds and enhance survival strategies aided by other species, and this is only what we see in the frequency spectrums that we normally perceive. Who knows what might be happening in the spectrums that other species can perceive that we cannot?

In the vegetable kingdom, as shown in the complex interaction of our olfactory and taste receptors, these characteristics combine to help us identify what is healthy to ingest and aid us.

On vision quests and in visionary states, particularly those encountered on shamanic plant diets centered around Ayahuasca, a pan-Amazonian psychoactive brew used among the indigenous people of the Amazon basin, subjects imbibe combinations of specialized plants. This is done in these traditions to learn from the spirits of the plants by passing through physical, mental, and spiritual ordeals to prove that the participants are worthy of the gift of knowledge that the plant spirits have to give them.

Each plant and animal has its own spirit, essence, or energy characterized as its own personality in the same way that Ayahuasca is universally referred to as "The Mother." It is also the reason why North American Indians don't refer to animals as *the* bear or *the* coyote. They say Bear or Coyote as they consider them all to be manifesting the essence of their spirit in a grouped manner, similar to bees, ants, and other cooperative colonies, and each one has its own unique energy signature that is treated with equal respect, like the well-known personality of Coyote the Trickster.

When you ingest a particular plant or plants, especially in a specific environment while on a strict cleansing diet, their effects are strongly felt and enhanced, even more so the longer the diet is continued. During that time participants are subject to the energy field, or spirit of that plant or plants, which all interact with each other and the bodies and minds of the subjects in different ways.

The combination of the diet, environment, plants, and other elements affect brain wave activities and physiologies in a myriad of mysterious ways on physical, psychological, and spiritual levels that we can scarcely comprehend, much less study in any traditional, objective scientific way. The best way to study them is from the outside in through the subjective perceptions they produce in experiences that transpire on deep non-rational levels of resonance and synchronization. In these states the plants act as transmitters in the position of orchestra conductors, communicating in ways that defy logic and articulation, and the subjects are receivers.

An Ayahuasca ceremony is characterized as an individual experience in a group setting with much anecdotal evidence supporting sporadic telepathic experiences among participants. The healing circle is referred to as a container to safely hold the energy it generates and attracts so that the participants can be protected while in the vulnerable state necessary for healing.

If a group of people all sat with their own radios tuned to the same station listening to the same song, all of the radios would play the same music as one, in sync. If all of the participants in the healing circle tuned in and resonated at the same frequency or frequencies, there is no reason why they would not have similar shared telepathic events.

This learned ability to subject themselves to diverse natural energies to surrender and discover what mysteries they might reveal puts shamans in the humbling position of being subject to their plant teachers as conductor transmitters, in spite of the physical or psychological discomfort they may experience in the process.

In the Amazon, Ayahuasca is often referred to as *la purga*, because it purges the body and rigorously clears it out through every orifice, including vomiting, defecation, profuse sweating, tear ducts, and the lungs as well as energetically at deep psychological and spiritual levels. Often it is used in conjunction with other intensely purgative plants along with a restricted diet referred to as the *dieta* consisting of boiled rice, oatmeal, or quinoa, baked or boiled unripe bananas, and chicken or fish once a day or less combined with a pitcher a day of a helper plant or plants. There is no salt, soap, shampoo, toothpaste, scents or additions of any kind, and no sex. In addition to the deep inner purification, other plants are used for daily plant baths to physically and energetically wash away the toxins and energies that are released in the process.

These restrictions enhance the cleansing process over the time of the *dieta*, which dates back to prehistory. As time passes subjects lose their telltale human scents, particularly pheromones, hormones, and other secretions that jaguars and other animals with a highly refined sense of smell can detect, making the hunters virtually invisible to them and their primary sensory modalities. The constant refined ingestions and plant baths result in the subject not only smelling like the jungle, but they *become* the jungle, giving them a distinct hunting advantage. These practices fine tune the subject's own senses bringing great clarity and a highly refined perception of their immediate reality, further adding to their hunting advantage.

Aside from the intensified state of awareness that comes from this extreme purification which historically prepared them for hunts, battles, or other challenges, their conscious awareness of their field of perception expands at inner and outer levels making them psychologically and mentally clear for the task at hand.

These hard won physical and perceptual enhancements bring deep inner shifts from the resonance that comes from being sympathetic to and in sync with the plant spirits and the energies they manifest. One of the intriguing results of this immersion into what shamans characterize as the spirit world is the agreed upon encompassing energetic field that opens up the ability to not only commune with the plants and the distinctive energies of their unique spirits and personalities, but also to the distinctive energies and personalities of the animal kingdom.

This phenomenon is reinforced by widespread reports from Ayahuasca drinkers of directly experiencing and communing with or being "possessed" by specific animal energies that are common to the Ayahuasca experience, regardless of whether they are in the Amazonian jungle or in a major North American city.

Among the numerous totems claimed by participants, or which they say actually choose them, the most common are condors, jaguars, and snakes which have a deeper meaning going all the way back to prehistoric cultures. Those experiencing these energies or spirits often roar and growl through no volition of their own like jaguars, flap their legs like wings, or feel their bodies swaying seemingly of its own accord to distinctive serpentine movements. Other animals and insects like hummingbirds, butterflies, dragonflies, dolphins, and other aquatic totems can play big parts as well.

This phenomenon with its roars, "wing flapping" and other indicators that an animal spirit is present or "possessing" its host forms the core of shape shifting mythologies where shamans reportedly have the ability to change their form into that of their familiar animal totems.

In this way shamans share the distinct, intimate resonant vibration of these entities in the same energetic way they do with the morphogenetic field of plants, becoming one with them by sharing the same frequency. In the same way they have to surrender to the plant teachers to discover what mysteries may be revealed, shamans must humble themselves to their animal familiars as the conductor transmitters of their experience.

In the lore of the jungle, by sharing in the vibratory field of the animal's spirit energy, in a state of surrender on that entity's terms, that animal learns from the human by seeing things through human eyes while the human learns other modes of perception from seeing things through the animal's eyes. By mutual agreement and the understanding that comes from sympathetically sharing the same energetic field and perception of specific animals, combined with the invisibility that comes from the physical purifications that makes the hunters virtually invisible to the sensory modalities of their prey, the shaman hunts the animals in spirit first, then follows through what has happened in the spirit world in the actual physical hunt, aided by his cultivated invisibility.

This learned ability of shamans to tune in to and commune with these diverse plant and animal spirits allows them to experience these energies in a direct, definitive, subjective manner that helps them understand other perspectives. This cultivates empathy which is the ability to understand and share the feelings of another, and opens them up in ways that only direct experience can, while aiding them in developing the critical skills necessary for what is defined as soul retrieval.

Anything that comes into perception from an external objective source acts upon the subject creating an interference pattern where awareness takes place, defining the old adage "Where attention goes, energy flows," and like the wave/particle paradox, the opposite also holds true, "Where energy flows, attention goes."

THE GOLDEN MEAN IN THE MORPHOGENETIC FIELDS OF PLANTS

Plants are complex structures that change their shape in response to environmental factors like sunlight, water, and neighboring plants. Mathematical rules can be found in their growth patterns, particularly the golden section which is found in branching systems, phyllotaxis, flowers, seeds, and the spiral arrangement of plant organs. The golden section can also be found in the bifurcate angles of trees and lobed leaves.

Although trees and bushes differ in shape, their ratio of length to width is close to the golden section. In some plant stems, the divergence angle between two adjacent leaves approximates 137.28 degrees, the central angle forming two radii, and if we divide the circumference into two parts, the ratio is 1:0.618. This angle promotes adequate ventilation of the plants and is the optimal arrangement for light absorbance. Numbers of leaves, petals, and fruiting bodies in some plants also follow the sequence of Fibonacci numbers as is the case with spiral phyllotaxis, or the layout of seeds in pine cones, sunflowers, and other plants.

Fibonacci Spirals in a Pine Cone

If you count the spirals in a sunflower, there are **55** with either **34** or **89** on either side going in a counter clockwise direction, which is a Fibonacci sequence. If a Fibonacci number is divided by its immediate predecessor in the sequence, the quotient approximates the golden mean.

Fibonacci Spirals in a Sunflower

Fractals in a Romanesco broccoli

For most bushes or grasses like bamboo, the bifurcation angle is 34.4 degrees. For trees, poplar is 34.4 degrees, and peach is 55.6 degrees. These angles of 55.6 degrees and 34.4 degrees are the golden section of 90 degrees.

The golden section also exists in some lobed leaves like the lobe angles of Korean arborvitae and cypress which are both 34.4 degrees. The ratio between two pine needles is 0.618, as well as the ratio of leaf venation. This proportion is almost omnipresent, and an essential feature of plants which has great significance in biology and is a common phenomenon within the growth and evolutionary processes of plants.

Plant branch structure is complex, and there is a large diversity of forms among different species, but plant branching systems can be defined as two major kinds. Single-axis patterns like poplar and pine trees result in the main stem being of a higher status with respect to growth, while the lateral branches are inferior, making the plant shape high and straight. Mature trees have six or more branches and growth direction. The angle between the two neighbors is about 135 degrees and the angle between the main stem and each branch is close to 34.4 degrees which is the golden section of 90 degrees. Within 20 poplars, for every 100 branches, 79 branches satisfy this rule which is advantageous for leaves to absorb light and optimize the rate of photosynthesis.

Alternatively there is the Gather-axis pattern seen in camphor and pear trees where the terminal bud of the branch dies within a certain period of time and the lateral buds displace them forming strong side scions off the main axis. This pattern does not have obvious main branches, but forms many main stems causing the plant to have a sphere-like shape. A mature tree has 2–5 branches, and the angle between branches and the main stem is generally 55.6 degrees. In 20 camphor trees, for every 100 branches, 58 branches are near to this angle.

Leaves vary in the form of the leaf blade, lobes, edge and venation, many of which are related to the golden section. In Korean arborvitae, each leaf has six vein branches and some have five, while the scale-like shape is derived from branches that create a flat surface. The leaf has two growth directions; the angle between the main venation and branches is about 34.4 degrees. Among 100 Korean arborvitae leaves, 92 leaves satisfy this golden section. In Chinese pines, needles are grouped in twos. One is short, the other is long and the ratio of the two needles is 0.618. The angle between needles and the stem is near to 55.6 degrees. Among 100 needle pairs, 84 pairs satisfy this golden section. For some lobed leaves like those on maple and phoenix trees, the lobe angle is roughly 55.6 degrees. Blade leaves like those on poplar and camphor trees have an ellipse shape with a ratio between the short axis and the long axis of about 0.618.

This arrangements of leaves in some plants is referred to as phyllotaxis are typically Fibonacci numbers. The ratios of alternate Fibonacci numbers are given by the convergents to Φ^{-2}, where Φ is the golden section. This closely fits the fraction of a turn between successive leaves on the stalk of certain plants: 1/2 for elm and linden, 1/3 for beech and hazel, 2/5 for oak and apple, 3/8 for poplar and rose, 5/13 for willow and almond, and so on.

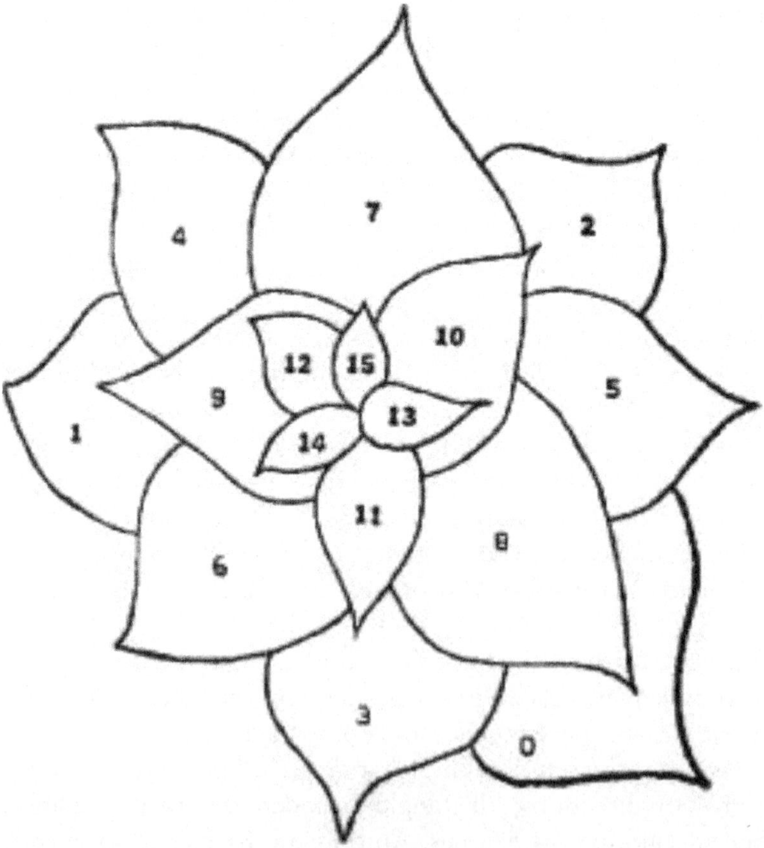

Phyllotaxis

Lilies, irises, and trillium have three petals; columbines, buttercups, larkspur, and wild rose have five petals; delphiniums, bloodroot, and cosmos have eight petals; corn marigolds have 13 petals; asters have 21 petals; and daisies have 34, 55, or 89 petals, all of which are Fibonacci numbers.

Some flowers have a large number of petals like the Chinese rose and are often arranged in spiral patterns with delamination. From inside to outside, the numbers of petals are 21, 13, 8, 5, and 3 based on their size, and these numbers are adjacent Fibonacci numbers.

The phyllotaxis of a Chinese Rose

Aside from their literal common root in the form of the divine proportion and all of its multifaceted permutations, all life on the planet earth is interconnected and works together in various and inventive ways. These arrangements show the cleverness of adaptation, diversity, and cooperation which are dependent on each other, allowing them to coexist in the world in a complex web of interactions, mutual aid, and interdependency.

THE SYMBIOSIS BETWEEN HUMANS, PLANTS, ANIMALS, AND OTHER ORGANISMS

A prime example of symbiosis is the inverse relationship between humans and plants pointed out by Rudolf Steiner, the founder of the Waldorf Schools.

Both plants and humans grow upright and the stem of the plant is like the spine of the human, although the stem is on the outside and the spine is on the inside. The stem is hollow and carries water and air to parts of the plant while the spine is hard and made of bone. Both structures support the plant or human being to stand upright and both are flexible.

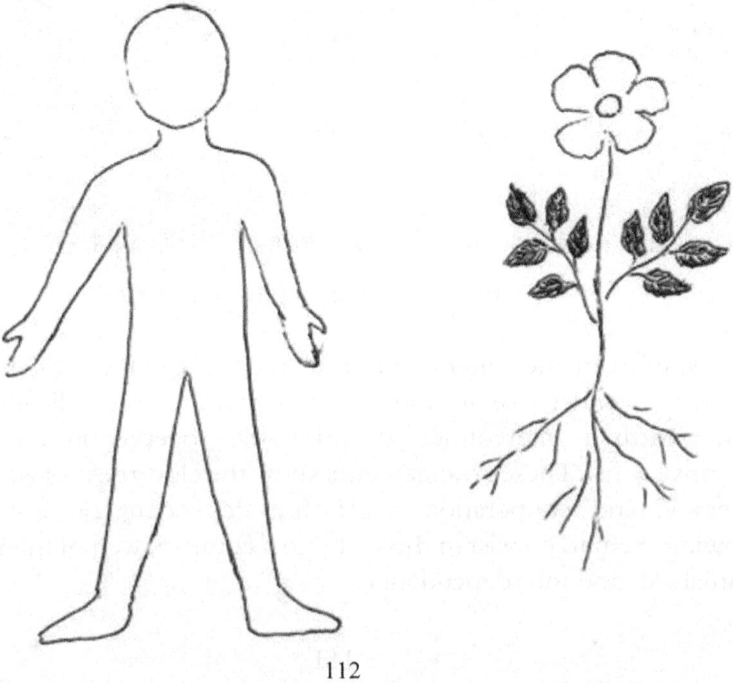

The leaves of a plant are like hands and the shape of a hand is similar to a leaf which makes food for the plant. Our hands feed us but can't make food like plants which make food inside themselves, while humans prepare food outside of themselves.

The roots of a plant grow out in all directions underground from their bottom to hold it in the soil where they drink up water and minerals from the soil, while humans take in food and water from their head through the mouth at the top. Like the roots at the bottom going out in all directions, from our head at the top our eyes look out everywhere around us, while our ears listen to our surroundings and our noses smell everything around us.

We speak about "flower heads" or the "pretty faces" of flowers that attract insects to pollinate them so they can make seeds for new plants. In another inverse relationship the places for the reproductive areas of the human being are hidden away, while the place for the plant to reproduce is out in the open.

If we examine a picture of a human with a plant upside-down inside of them, the head and the roots are in the same place, which is the coolest place, and the flower is in the warmest place where reproduction occurs.

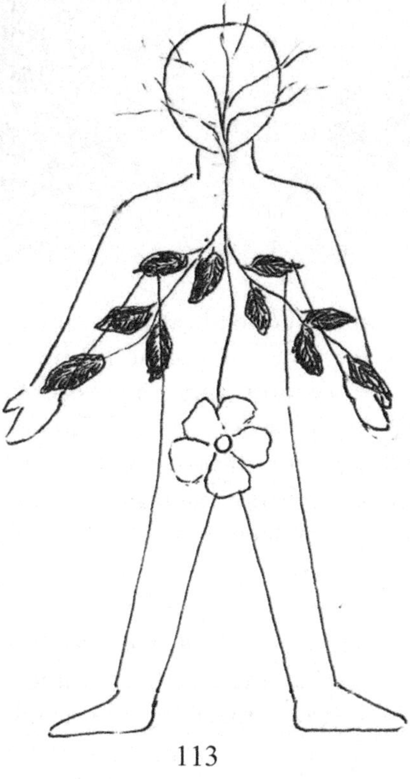

Plants also breathe in an opposite manner of the way humans do. Humans breathe in oxygen and breathe out carbon dioxide while plants breathe in carbon dioxide and breathe out oxygen. If you turn the respiration supplying branches of a tree inside out they are the identical inverse of human lungs, the human organ of respiration.

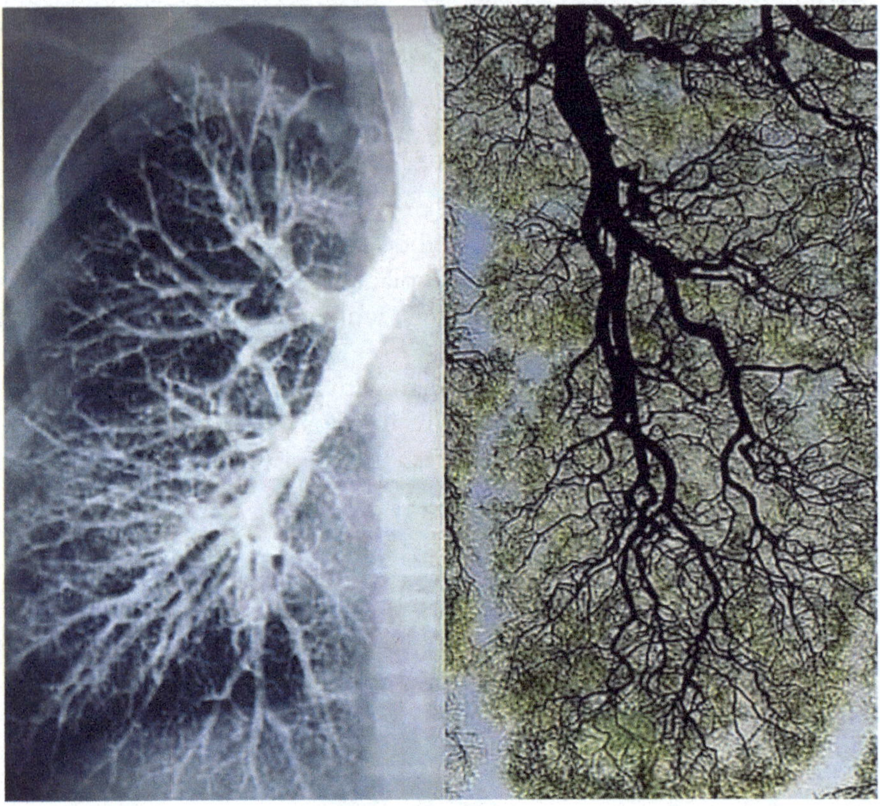

The fragrances that exist outside of plants are similar to our feelings. Steiner said that the fragrance of a plant is the most spiritual part of it. People who are good, kind, and compassionate to others develop beauty of the heart, which exists at the center of their being. In the same inverse way that plants show us beauty on the outside, evolved humans show their beauty from the inside.

The human/plant connection is a classic example of symbiosis, which is defined as any type of a close, long-term biological interaction between two different biological organisms. These organisms, each termed a symbiont, have to be different species and their interaction can be mutualistic, commensalistic, or parasitic. Symbiosis can be obligatory, meaning one or more of the symbionts depend on each other for survival like the example of humans and plants, or facultative, or optional, when they can live independently.

Symbiosis is also classified by physical attachment. When symbionts form a single body it is called conjunctive symbiosis, while all other arrangements are called disjunctive symbiosis. When one organism lives on the surface of another like head lice on humans, it is called ectosymbiosis, and when one partner lives inside the tissues of another like *Symbiodinium* within coral, it is termed endosymbiosis.

These relationships can be obligate, meaning one or both of the symbionts depend on each other for survival like in the case of lichens, which consist of fungal and photosynthetic symbionts where the fungal partners cannot live on their own. The algal or cyanobacterial symbionts in lichens like *Trentepohlia* can live independently, and their part of the relationship is described as facultative or non-obligate.

Endosymbiosis is any symbiotic relationship where one symbiont lives within the tissues of the other, either within the cells or extracellularly. Examples include diverse microbiomes like rhizobia, which are nitrogen-fixing bacteria that live in root nodules on legume roots; actinomycetes, nitrogen-fixing bacteria like Frankia that live in alder root nodules; single-celled algae inside reef-building corals; and bacterial endosymbionts that provide essential nutrients to about 10–15 percent of insects.

Ectosymbiosis is any symbiotic relationship where the symbiont lives on the body surface of the host, including the inner surface of the digestive tract or the ducts of exocrine glands. Examples of this include ectoparasites like lice and commensal ectosymbionts like barnacles that attach themselves to the jaw of baleen whales; and mutualist ectosymbionts like cleaner fish.

Mutualism or interspecies reciprocal altruism is a long-term relationship between individuals of different species where both individuals benefit. Mutualistic relationships may be either obligate for both species, or obligate for one, but facultative for the other, or facultative for both.

A large percentage of herbivores have mutualistic gut flora made up of cellulose-digesting protozoans or bacteria living in the herbivores' intestines to help them digest plant matter, which is more difficult to digest than animal prey.

Coral reefs are the result of mutualism between coral organisms and various types of algae that live inside them. Most land plants and land ecosystems rely on mutualism between the plants that fix carbon from the air, and mycorrhizal fungi that help extract water and minerals from the ground.

An example of mutualism is the relationship between the ocellaris clownfish that dwell among the tentacles of Ritteri sea anemones. These territorial fish protect the anemone from anemone-eating fish, and the stinging tentacles of the anemone protect the clownfish from its predators. A special mucus on the clownfish protects it from the stinging tentacles.

Another example is the goby, a fish that lives together with shrimp. The shrimp digs and cleans a burrow in the sand where the shrimp and the goby fish live. The shrimp is almost blind, leaving it vulnerable to predators when outside its burrow. In case of danger the goby touches the shrimp with its tail to warn it. When that happens both the shrimp and goby retreat into the burrow. Different species of gobies also clean up ectoparasites in other fish.

A facultative symbiosis is seen in encrusting bryozoans and hermit crabs. The bryozoan colony develops a cirumrotatory growth and offers the crab a helicospiral-tubular extension of its living chamber that initially was situated within a gastropod shell.

Many types of tropical and sub-tropical ants have also evolved complex relationships with certain tree species. Azteca ants and Cecropia trees have a symbiotic relationship with the ants using the trees as their homes. When holes were drilled into Cecropia tree trunks in Panama, ants emerged from their homes to patch up the wounds, significantly reducing the size of the holes within 2 1/2 hours, leaving them totally healed within 24 hours. The tree's trunk is like an ant apartment with more floors added as the tree grows. Inside there are passages allowing the ants to move around with small openings letting them go outside. In exchange for their home, the ants defend the Cecropia leaves from herbivores.

In endosymbiosis, the host cell lacks some of the nutrients that the endosymbiont provides. As a result the host favors endosymbiont growth processes within itself by producing specialized cells that affect the genetic composition of the host to regulate the increasing population of the endosymbionts and ensure that these genetic changes are passed on to the offspring.

A spectacular example of obligate mutualism is the relationship between siboglinid tube worms and symbiotic bacteria that live at hydrothermal vents and cold seeps. The worm has no digestive tract and is wholly reliant on its internal symbionts for nutrition. The bacteria oxidize either hydrogen sulfide or methane supplied by the host to them.

Commensalism describes a relationship between two living organisms where one benefits and the other is not significantly harmed or helped.

Commensal relationships can involve one organism using another for transportation, or for housing, or it can involve one organism using something another created after its death. Examples of this relationship are hermit crabs using gastropod shells to protect their bodies, and spiders building their webs on plants.

In parasitic relationships, the parasite benefits while the host is harmed. Parasitism takes many forms, from endoparasites that live within the host's body to ectoparasites and parasitic castrators that live on its surface, and micropredators like mosquitoes that visit intermittently. Parasitism is an extremely successful mode of life. About 40 percent of all animal species are parasites, and the average mammal species is host to 4 nematodes, 2 cestodes, and 2 trematodes. Their predatory nature plays its part within the interconnected web of everything else to help control populations and keep the natural order in balance.

Mimicry is a form of symbiosis where a species adopts characteristics of another species to alter its relationship with the species being mimicked to its own advantage. Among the many types of mimicry are Batesian and Müllerian, the first involving one-sided exploitation, the second providing mutual benefit. Batesian mimicry is an exploitative three-party interaction where one species, the mimic, has evolved to mimic another to deceive a third, known as the dupe. In terms of signaling theory, the mimic and model have evolved to send a signal; the dupe has evolved to receive it from the model. This is to the advantage of the mimic but to the detriment of both the model, whose protective signals are effectively weakened, and the dupe, which is deprived of an edible prey. As an example, a wasp is a strongly-defended model, which signals with its conspicuous black and yellow coloration that it is an unprofitable prey to predators like birds that hunt by sight. Many hoverflies are Batesian mimics of wasps, and any bird that avoids these hoverflies is a dupe.

In contrast, Müllerian mimicry is mutually beneficial as all participants are both models and mimics. An example of this is when different species of bumblebee mimic each other with similar warning coloration in combinations of black, white, red, and yellow, and all of them benefit from the relationship.

Cleaning symbiosis is an association between individuals of two species, where one, the cleaner removes and eats parasites and other materials from the surface of the other, the client. This relationship is mutually beneficial, but biologists have debated whether it is mutual selfishness, or simply exploitative. Cleaning symbiosis is well known among marine fish where some small species of cleaner fish are specialized to feed almost exclusively by cleaning larger fish and other marine animals.

Flowering plants and the animals that pollinate them have also co-evolved. Many plants that are pollinated by insects, bats, or birds have highly specialized flowers modified to promote pollination by a specific pollinator that is correspondingly adapted.

Hummingbirds are specialized nectarivores and are tied to the bird pollinated flowers they feed on. Some species, especially those with unusual beak shapes like the sword-billed hummingbird and the sicklebills co-evolved with a small number of flower species, but even in the most specialized hummingbird-plant relationships the number of food plant lineages of individual hummingbird species increases with time. The bee hummingbird which is the world's smallest bird evolved to dwarfism most likely because it had to compete with long-billed hummingbirds having an advantage for nectar foraging from specialized flowers, leading the bee hummingbird to compete more successfully for flower foraging against insects.

Many plants pollinated by hummingbirds produce flowers in shades of red, orange, and bright pink, though the birds will take nectar from flowers of other colors. Hummingbirds can see wavelengths into the near-ultraviolet, but hummingbird-pollinated flowers do not reflect these wavelengths like insect-pollinated flowers do. This narrow color spectrum may render hummingbird-pollinated flowers inconspicuous to most insects, reducing nectar robbing. Hummingbird-pollinated flowers also produce weak nectar, while insect-pollinated flowers typically produce more concentrated nectars dominated by fructose and glucose.

The first flowering plants in the fossil record had simpler flowers until adaptive speciation gave rise to diverse groups of plants. At the same time, corresponding speciation occurred in certain insect groups. Some groups of plants developed nectar and large sticky pollen while insects evolved more specialized morphologies to access and collect these rich food sources. In some taxa of plants and insects the relationship has become dependent, where the plant species can only be pollinated by one species of insect.

The acacia ant is an obligate plant ant that protects at least five species of acacia from preying insects and other plants competing for sunlight, while the tree provides nourishment and shelter for the ant and its larvae.

Seed dispersal is the movement, spread or transport of seeds away from a parent plant. Plants have limited mobility and rely on a variety of dispersal vectors like the wind and living biotic vectors like birds. In order to attract animals, these plants evolved a set of morphological characteristics like fruit color, mass, and persistence correlated to particular seed dispersal agents. As an example, some plants evolve conspicuous fruit colors to attract avian frugivores and birds learn to associate these colors with a food resource.

Symbiosis in all of its forms interact within a far more sophisticated web of life, all of which contributes to a delicate homeostasis that keeps the planet earth thriving in the same manner as the complex organism known as the human body until these processes are interrupted by forces both human and natural that undermine its precision and contribute to its destruction.

SYMBIOSIS AND THE LANGUAGE(S) OF PLANT TALK

Within the soil beneath our feet is an ecosystem as diverse as a rainforest or a coral reef that we depend on for 99 percent of our food. Typically, several hundred thousand small animals live under one square meter of undisturbed ground in the Earth's mid-latitudes and 90 percent of the species they belong to have yet to be named. Less than a teaspoonful of this soil contains around a kilometer of fungal filaments.

Springtails, which are tiny animals similar to insects thrive in dozens of shapes and sizes, and round, crabby mites are everywhere, often in numbers up to half a million in every square meter. Among them are other life forms that look like tiny white centipedes turned out, called symphylids, and long low creatures with two fine antennae at the front and two at the back, looking like a dragon or a flying horse, called a bristletail, or dipluran.

That this thin cushion between rock and air can withstand all we throw at it and still support us is a dangerous belief. Most people see the soil as a dull mass of ground-up rock and dead plants, but in reality it is a complex biological structure built by living creatures to secure their survival, like a wasps' nest or a beaver dam. Microbes make cement out of carbon that they stick mineral particles together with creating pores and passages that allow water, oxygen, and nutrients pass. The tiny clumps they build become the blocks the animals in the soil use to construct bigger labyrinths.

Soil is fractally scaled, meaning its structure is consistent, regardless of magnification. Bacteria, fungi, plants and soil animals working together building an immeasurably intricate, endlessly ramifying architectures that organizes itself into coherent worlds which contribute to soil's resistance to droughts and floods.

It also reveals why soil can break down so quickly when it's farmed. Under certain conditions, when farmers apply nitrogen fertilizer, microbes respond by burning through the carbon that holds their catacombs together, then the pores cave in, the passages collapse, and the soil becomes sodden, airless, and compacted.

Plants release between 11 and 40 percent of all the sugars they make through photosynthesis into the soil and deliberately pump them into the ground. Before releasing them, they turn some of these sugars into compounds of tremendous complexity.

These complex chemicals are pumped into the zone surrounding the plant's roots called the rhizosphere where they are released to create and manage its relationships.

Soil is full of bacteria and its earthy scent comes from the compounds they produce. Most of the time these bacteria wait in suspended animation for messages that wake them in the form of complex chemicals the plant releases that alerts the particular bacteria to promote its growth. Plants use a sophisticated chemical language that only the microbes that they want to speak with can understand.

When a plant root pushes into a lump of soil it releases messages that trigger an explosion of activity. The bacteria responding to its call consume the sugars the plant feeds them and proliferate to form some of the densest microbial communities on Earth. There can be a billion bacteria in a single gram of the rhizosphere that unlock the nutrients the plant depends on and produces growth hormones and other chemicals that help it grow.

The plant's vocabulary changes from place to place and time to time, depending on what it needs. If it's starved of certain nutrients, or the soil is too dry or salty, it calls out to the bacteria species that can help.

Although the rhizosphere lies outside the plant, it functions as if it were part of the whole and can be thought of as the plant's external gut. The similarities between the rhizosphere and the human gut where bacteria also live in astonishing numbers are uncanny. In both systems microbes break down organic material into simpler compounds that the plant or person can absorb. Though there are more than a thousand major groups of bacteria, the same four dominate both the rhizosphere and the guts of mammals.

In the same way that human breast milk contains sugars called oligosaccharides whose purpose is not to feed the baby, but the bacteria in the baby's gut, young plants release large quantities of sucrose into the soil to feed and develop their new microbiomes, and just as the bacteria that live in our guts outcompete and attack invading pathogens, the friendly microbes in the rhizosphere create a defensive ring around the root. Additionally the same way bacteria in the colon educate our immune cells and send chemical messages that trigger our body's defensive systems, the plant's immune system is trained and primed by bacteria in the rhizosphere.

In conjunction with this intimate communication with other species within their immediate environment, trees talk to each other deep underground using a complex, highly evolved symbiotic network that is identical to a human neural network that functions as a conducting circuit of biological neurons. Through fungal highways that connect them, the oldest trees nurture their young and communicate and cooperate with other species.

Aside from trees using this complex fungal network for communicating amongst themselves, studies show that fungi also "talk" to each other independently of their interaction with trees.

Professor Andrew Adamatzky at the University of West England in Bristol published a study that described how oyster fungi generate electric pulses called "spikes" indicating that they were sharing information with each other.

"We do not know if there is a direct relationship between spiking patterns in fungi and human speech. Possibly not," Adamatzky said. "On the other hand, there are many similarities in information process in living substrates of different classes, families, and species. I was just curious to compare."

Adamatzky recorded the electrical activity of four species of fungi, resulting in the discovery that mushrooms can not only speak to each other, but they have a vocabulary of roughly 50 different "words".

"We found that spikes are also clustered," reads the study. "Assuming that spikes of electrical activity are used by fungi to communicate and process information in mycelium networks, we group spikes into words and provide a linguistic and information complexity analysis of the fungal spiking activity.

There is an emerging body of studies on language of creatures without a nervous system and invertebrates," explains the study, "…a modified conception of language of plants is considered to be a pathway towards the de-objectification of plants and the recognition of their subjectivity and inherent worth and dignity."

Amanita muscaria, also known as the fly agaric or fly amanita

Like mushrooms, plants and trees also "talk" to each other independently of their interaction with mycelial networks, and some aspects of plant behavior can be studied in the same terms as animal and human behavior.

Plants have nervous systems like animals and use some of the same compounds in communications like trimethylamine-N-oxide known as TMAO, a molecule associated with human heart disease and marine animals' ability to survive high-pressure conditions. TMAO helps plants cope with stressful conditions the same way it works in animals.

Plants also use glutamate, which is the most abundant free amino acid in the brain to speed transmission. Animals require rapid, long-range molecular signaling networks to integrate sensing and response throughout their bodies. Glutamate acts as an excitatory neurotransmitter in the vertebrate central nervous system, facilitating long-range information exchange via activation of glutamate receptor channels. Similarly, plants sense local signals like an herbivore attack, and transmit this information throughout the plant body to rapidly activate defense responses in undamaged parts.

Claude Bernard, a French physiologist, demonstrated that the shameplant classified as Mimosa pudica which folds into itself when touched, was unresponsive to touch after exposure to ether, a commonly used anesthetic. The shameplant also folds into itself at night, but this movement was not affected by anesthesia. Bernard concluded that anesthesia does not inhibit the ability to move, but it inhibits the plant's ability to sense its environment, indicating that anesthesia blocks consciousness.

According to plant physiology expert Rainer Hedrich, Venus flytraps remember when they are touched. When prey lands on the plant's trap, it brushes against a sensory hair. The hair fires an electrical impulse and releases a wave of signal molecules throughout the trap. After two impulses, the trap closes and imprisons the animal prey. After five impulses, the plant produces digestive enzymes. Because anesthesia disrupts memory in animals, Hedrich hypothesized that anesthesia prevented the plant from remembering each stimulation.

To test this, Hedrich determined if anesthetized flytraps still release signaling molecules. They found that sensory hair still released the signaling molecules when stimulated, but the signal did not spread throughout the trap. In an animal, this is similar to local pain receptors detecting pain and releasing local pain signals without those signals ever reaching the brain. The Venus flytrap's response to anesthesia suggests that anesthesia affects plants at cellular and organ levels similar to animals.

Plants also use RNA to "talk to" neighbors, affecting their gene expression. Simple responses to the facts of nature aid the plant in learning; for example, response to the position of the sun, response to recent temperature changes, losing or gaining a response to a stimulus, and associating one stimulus with another. Stimulus response and its diverse associations are also studied in animal and human psychology.

In terms of plant memory, during the day, the mallow plant uses motor tissue at the base of its stalks to turn its leaves toward the sun, a process that's actively controlled by changes in water pressure inside the plant. The magnitude and direction of the sunlight is encoded in light-sensitive tissue spread over the mallow's geometric arrangement of leaf veins and stored overnight. The plant also tracks information about the cycle of day and night via its internal circadian clocks which are sensitive to environmental cues that signal dawn and dusk.

Overnight, using information from all these sources, the mallow can predict where and when the sun will rise the next day. It might not have concepts like "the sun" or "sunrise", but it stores information about the light vector and day/night cycles that allow it to reorient its leaves before dawn so that their surfaces face the sun as it climbs in the sky. This also allows it time to re-learn a new location when plant physiologists mess with its "head" by changing the direction of the light source. When the plants are shut in the dark, the anticipatory mechanism also works offline for a few days, optimizing available resources, in this case, sunlight.

Plants store information about their experiences like the Venus flytrap which can time things by the chemicals circulation in their systems. Although a mind is required to work with ideas, many natural systems store information by other means.

Venus flytrap with trapped prey

After millions of years of evolution, fungi and plants formed symbiotic relationships called a mycorrhiza, a word that comes from the Greek meaning fungus and root. In exchange for sugars and carbon from the trees, fungi provide minerals, nutrients, and a communication network that trees rely on. Similar to an internet connection, the mycorrhizal network extends throughout the forest. Fungal threads called hyphae create a highway and merge with tree roots allowing trees to send and receive items like nitrogen, sugars, carbon, phosphorous, water, defense signals, chemicals, and hormones. One tree can connect to hundreds of other trees, sending out signals, and along these threads, bacteria and other microbes swap nutrients with the fungi and tree roots.

This underground world of infinite biological pathways connects trees, allowing the forest to behave as a single organism with its own inherent intelligence. Reaching out along the network, hub trees, called Mother Trees nurture growing saplings. When older trees die they bequeath their nutrients, genes, accumulated wisdom to others. By tapping into the otherworld trees gain valuable resources and insight into their surroundings.

As a consequence connected trees have a distinct advantage and resilience, but if you cut a tree off from the network it becomes vulnerable and often succumb to disease at higher rates. Practices like clear-cutting or replacing forests with a single species decimates this intricate ecosystem and harvesting becomes unsustainable.

THE TREES OF LIFE

Most human cultures that revered ancient trees see them as bridges connecting the divine and the human and they are considered sources of knowledge, life, fertility, and longevity. Abusing, disrespecting, or misusing one of these trees can unleash disaster and bring death, as in the biblical tale of the Garden of Eden, with its tree of life and of knowledge of good and evil. The trees in these stories are always different from other trees in the forest or the Garden and are seen as belonging to a separate, special, transcendent class of beings.

A study, co-authored by three scholars specializing in biological and environmental sciences and evolutionary biology, examined how these ancient trees play a key role in the development and maintenance of forests all over the world. Referred to as "Trees of Life", these exceptional trees are usually ten to twenty times older than the average plant species in their ecosystems.

These scientists discovered that ancient trees sustain the health of the surrounding trees, keep forests from dying out, and keep them thriving for millennia. By examining the demographic patterns of ancient forests they discovered that fewer than 1 percent of the trees in the environment "emerge as life-history "lottery winners" that reach higher ages that bridge environmental cycles that span centuries. "In our models," Dr. Chuck Cannon explained, "these rare, ancient trees prove to be vital to a forest's long-term adaptive capacity, substantially broadening the temporal span of the population's overall genetic diversity."

Despite their extreme age, the study finds these "Trees of Life" have survived countless environmental changes. This resilience allowed them to pass on their genetic material to the rest of the organisms sprouting around them, while cutting off a large amount of carbon in comparison to the typical old tree in the woods.

Trees have been revered as sacred monuments since the prehistoric era. Our ancestors were inspired by their annual cycle of decay in the autumn followed by a luscious rebirth in the spring and to the primitive mind these trees became symbols of life, death, and rebirth.

The Tree of Life shown here in its geometrical representation in association with the Flower of Life held mythical status throughout all world cultures as the World Tree. Its branches were said to reach as high as the heavens, while its roots plunged deep into the abyss of the underworld. Because of their association with celestial realms, these trees were regarded by many shamans and priests as gateways to other dimensions.

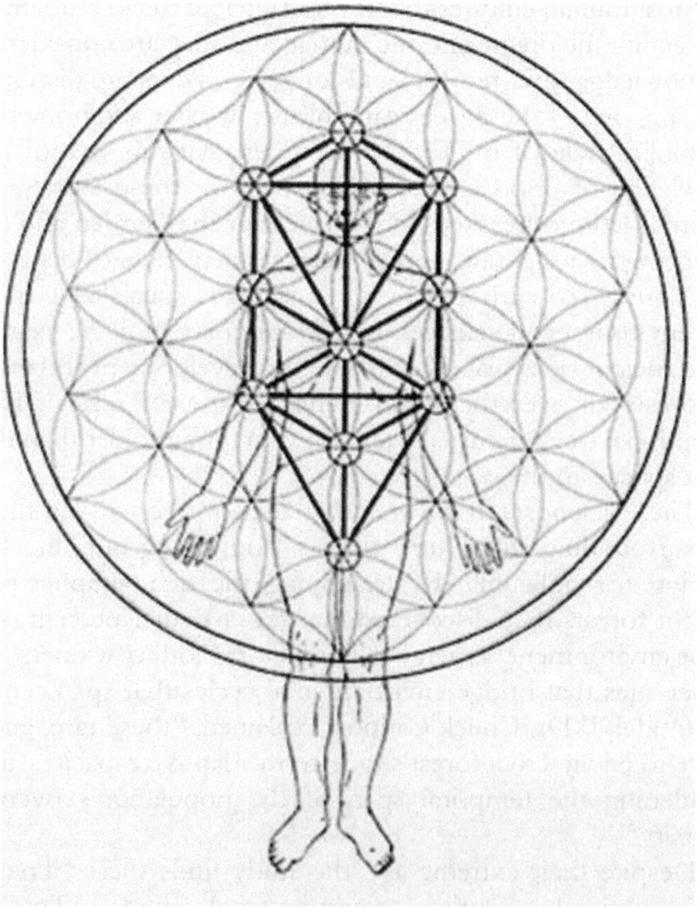

Tree of Life

• Yggdrasil was said to connect middle earth (Midgard) to eight other realms, some made of fire and ice, others of darkness and light.

• The "Sky Tree" from Hungarian mythology had a series of branches that reached out to seven worlds. Each branch was said to touch the sun, the moon, the clouds and other celestial spheres.

• The Cosmic Tree of Latvia was a highly symbolic image said to hold birds in its branches representing enlightenment, mammals at its base for instinct, and serpents in its roots for wisdom

• The World Tree of Mesoamerica was seen as a gateway known as the Axis Mundi that connects the planes of the Underworld and the sky with the terrestrial world.

• It was Adam and Eve who attained wisdom by eating fruit from the Tree of Knowledge in Judaism and Christianity. In many depictions of the World Tree, a great serpent is said to reside at its base guarding forbidden knowledge which only a select number of mortals have ascertained.

- The Druids were priests of the natural world whose name was said to mean "knower's of the oak" in Celtic mythology.

- The Dawn Tree from Baltic mythology was depicted with a golden trunk, copper roots and silver leaves. Its trunk represented life in the present while the past was embodied in its roots, and the branches represented choices yet to be made.

- The Kabbalah represents the tree of life which is the other legendary tree of Christian mythology. Legend says that to learn the secrets of this mystic tree is to know the secrets of life itself.

- In Tengrinism, it is believed that humans were descended from Trees, as well as other spirit entities and lesser gods who lived for hundreds of years.

- It was under the Bodhi tree that the Buddha was said to have gained enlightenment in Buddhist mythology.

- A tree of life belonged to the Goddess Iusaaset who was said to have conceived the lineage of Egyptian Gods through the life-giving properties of the trees.

- The Kalpavriksha tree from Vedic mythology was said to produce an abundance of life-giving fruit for those who wished for it.

The most common theme found in world mythology is the "Tree of Life" which is said to both create and prolong life. In many cases our ancestors believed that these great trees gave birth to the Gods and even to humanity itself. Many pilgrims have searched for this mythical tree throughout the ages, but only the worthiest have been able to find it.

In Greek mythology Gaia is the personification of the Earth and one of the Greek primordial deities. Gaia is the ancestral mother of all life. Beliefs and worship amongst neopagans regarding Gaia range from the belief that Gaia is the Earth, to the belief that she is the spiritual embodiment of the earth or the goddess of the Earth.

In modern times The Gaia hypothesis also known as the Gaia theory, Gaia paradigm, or the Gaia principle, proposes that living organisms interact with their inorganic surroundings on Earth to form a synergistic, self-regulating, complex system that helps to maintain and perpetuate the conditions for life on the planet.

Topics related to the hypothesis include how the biosphere and the evolution of organisms affect the stability of global temperature, salinity of seawater, atmospheric oxygen levels, the maintenance of a hydrosphere of liquid water, and other environmental variables that affect the habitability of Earth.

The Gaia hypothesis was initially criticized for being teleological and against the principles of natural selection, but later refinements aligned it with ideas from fields like Earth system science, biogeochemistry, and systems ecology. This hypotheses suggests that organisms co-evolve with their environment and influence their abiotic environment, and that environment in turn influences the biota by Darwinian process. A reduced version of the hypothesis states that the biota influence certain aspects of the abiotic world like temperature and atmosphere.

In some versions of Gaia philosophy all lifeforms are considered part of one single living planetary being called *Gaia*. In this view, the atmosphere, the seas and the terrestrial crust are the results of interventions carried out by Gaia through the coevolving diversity of living organisms.

The Gaia paradigm was an influence on the deep ecology movement, an environmental philosophy that promotes the inherent worth of all living beings regardless of their instrumental utility to human needs, and the restructuring of modern human societies in accordance with such ideas.

Deep ecology argues that the natural world is a complex of relationships where the existence of organisms is dependent on the existence of others within ecosystems and that non-vital human interference with or destruction of the natural world poses a threat not only to humans, but to all organisms constituting the natural order, something we are now experiencing first hand in modern times.

Deep ecology takes a holistic view of the world humans live in and seeks to apply to life the understanding that the separate parts of the ecosystem, including humans, function as a whole. Not only do they function as a whole, but the implicate/explicate order is reflected throughout all of these life forms on multiple levels, whether microscopic, human, plant, animal, or any number of other organisms.

THE EARTH AS AN INTELLIGENT ENTITY

A team of scientists published a paper in the *International Journal of Astrobiology* presenting the idea of planetary intelligence to describe what they term as the collective knowledge and cognition of an entire planet.

These scientists believe that the concept might help deal with global issues, including climate change and extraterrestrial life. They pointed to the evidence that underground networks of fungi can communicate to suggest that large-scale networks of life could form a vast invisible intelligence that profoundly alters the condition of the entire planet.

Humans are changing the Earth's environmental balance. It might be able to adapt because it is an "intelligent entity", but because of the climate, and to a greater extent the plastic crisis, we may be irrevocably changing the environmental balance.

Adam Frank, professor of physics at the University of Rochester and coauthor of the paper said, "We don't yet have the ability to communally respond in the best interests of the planet". The researchers believe that seeing things from this perspective can help humans understand their impact on the Earth and serve as a guide on how to improve it, and they believe it can aid in the search for extraterrestrial intelligence as well.

"We're saying the only technological civilizations we may ever see — the ones we should expect to see — are the ones that didn't kill themselves, meaning they must have reached the stage of a true planetary intelligence," Frank said. "That's the power of this line of inquiry. It unites what we need to know to survive the climate crisis with what might happen on any planet where life and intelligence evolve."

Conventionally, intelligence is seen as a property of individuals as well as a property of collectives. Examples include collective-decision-making by social insects, slime mold navigating mazes, and intelligent behavior of individual cells and viruses which are themselves a collective of chemical processes. Human intelligence derives from our social behavior which is currently global in its reach. These examples highlight the fact that intelligence, broadly construed, operates across diverse length and time scales.

An open question is whether or not intelligence can operate at the planetary scale, and if so, how a transition to planetary-scale intelligence might occur and whether or not it has already occurred, or is on our near-term horizon. Understanding the current state of intelligence on Earth and steering its future requires understanding of how human and technological systems are integrated, and how they might display collective intelligence at a planetary scale.

The appearance of intelligence in the evolution of life could represent one of a series of major transitions in a planet's history. From Earth's evolution, we see that once intelligence manifests in the form of a global technological civilization, it has the power to reshape a planet in profound ways. On the one hand, a civilization's technological and energy-harvesting capacities might help it engineer the world, creating new planetary-scale behaviors and functions that allow civilization to survive over longer timescales than would otherwise be possible. Conversely, those same technological capacities could lead the species and its global civilization into a mass suicide, as could be the case with nuclear weapons. They also have the possibility of driving the planet into new states of its coupled systems like the atmosphere and hydrosphere, which could truncate civilization's evolution through climate change.

The consideration of intelligence in astrobiological studies generally regard it as the property of one or more species that evolve on a planet and create a technological civilization as opposed to a collective property existing in groups of organisms or societies. By implication most astrobiologists do not view intelligence as a property of the biosphere that a "technosphere" might emerge from as an evolutionary stage of global intelligence.

The appearance of technological intelligence can be seen as a planetary scale transition, but not as something that happens *on a* planet but as something that happens *to a* planet. Most researchers agree that the correct scale to understand key aspects of life and evolution is planetary, as opposed to the traditional focus on individual species.

The concept of *planetary intelligence* can be useful for three distinct domains: Earth Systems and Exoplanet studies, Anthropocene and Sustainability studies, and the study of Technosignatures and the Search for Extraterrestrial Intelligence known as SETI. **The Anthropocene is defined as the period of time that human activities have had an environmental impact on the Earth and is regarded as constituting a distinct geological age.**

The concept of planetary intelligence holds promise in providing a framework for understanding possible paths of long-term inhabited planetary evolution that is both broad and deep. It can also help unite disparate perspectives into a single explanatory paradigm for the transitions in the Earth-system observed in the past with what we are experiencing now, and what we might experience in the future evolution of the Earth.

Planetary intelligence is the acquisition and application of collective knowledge operating at a planetary scale integrated into the function of coupled planetary systems. One evolving example would be the global response to the planetary-scale crisis of ozonosphere erosion by CFCs. Another could be a global response to the crisis of anthropogenic global warming. These examples involve a global coordinated response to a potential existential threat and are dependent on decision-making at the level of localized activities of individuals and governments.

A transition to global planetary intelligence should include intelligence that is more than the aggregate sum of the localized activities of life on smaller scales as is the case with properties existing at the scale of biospheres.

Technospheres are the aggregate planetary activity of technology and in their coupling to other planetary systems like geospheres that are not apparent in individual organisms and subsystems comprising a biosphere or technosphere. The cognitive activity of interest has to operate through feedback loops that are global in scale, coordination, and operation.

The concept of human computation includes examples where humans are computational elements in information processing systems in crowd-sourced activities like wiki editing or human-assisted AI. By defining planetary intelligence as cognitive activity in terms of knowledge that is only apparent at a global scale we can broaden our view of technological intelligence beyond species that can reason or build tools in the traditional sense. Terms like "knowledge" and "cognition" usually describe individuals, but they can also apply to planetary-scale processes.

There are four possible domains of planetary intelligence to explore the operation and effect of planetary intelligence and each relates to a different, but successive phase of planetary evolution.

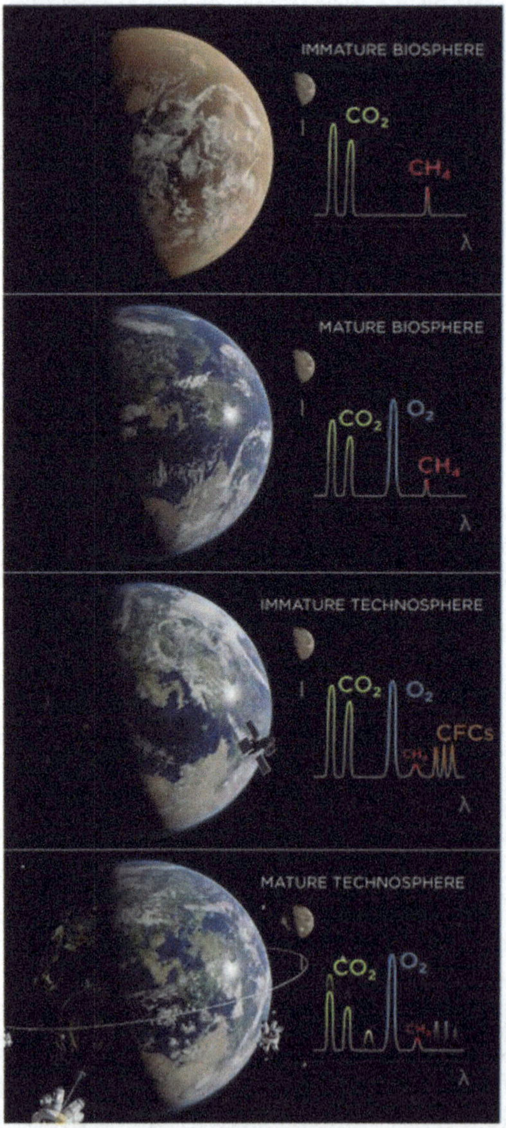

(a) On a planet with an immature biosphere such as the Earth during the Archean Eon there are insufficient feedback loops between life and geophysical coupled systems to exert strong co-evolution.

(b) On a planet with a mature biosphere like Earth after the Proterozoic, the biosphere exerts strong forcing on the geophysical state establishing full co-evolution of the entire system. This feedback can provide some long-term stabilizing modulations for the full system.

(c) On a planet with an immature Technosphere like the current Anthropocene Earth feedbacks from technological activity produce strong enough forcing on the coupled planetary system to drive it into new dynamical states, but these forcings are unconstrained by intention relative to the health of the civilization producing the technology.

(d) On a planet with a mature Technosphere, feedback loops between technological activity and biogeochemical and biogeophysical states have been intentionally modified to ensure maximum stability and productivity of the full system.

Alongside each image is the atmospheric spectrum.

An immature biosphere would show an atmosphere in equilibrium, but dominated, possibly by CO_2.

In a mature biosphere life has changed atmospheric chemistry leading to a non-equilibrium state with possible high concentrations of O_2.

In an immature Technosphere new "pollutant" species like CFCs appear, while industrial activities like combustion alters the abundance of other pre-existing gases like CO_2 and methane.

In a mature Technosphere all atmospheric constituents have their concentrations modified to produce long-term stable productive states for the full civilization and biosphere represented by a range of possible peaks for different constituents.

To consider intelligence or some form of cognition operating on a planetary scale on worlds without a planetary-scale technological species like item (a) and (b) would require some form of collective cognition to be a functional part of the biosphere for longer than the relatively short tenure of human intelligence on Earth. If true, then the global nature of the complex, networked feedbacks occurring in the biosphere could indicate the operation of an ancestral planetary intelligence.

To consider whether the changes humans have introduced to the planet through industrial activities like changes marking the "Anthropocene" geological epoch, it can be understood as a transition in both the kind and level of global cognitive activity through the emergence of networks of processes that originate with human agency. These would need to become active and autonomously operate on levels beyond individuals giving weight to the idea of an emerging technosphere and its place in the Anthropocene.

By focusing on the Anthropocene we can assess the sustainability requirements for a long-lived industrial planetary civilization through the lens of planetary intelligence. Many current threats to sustainability are characterized by planetary-scale changes in the environment caused by our combined activities being unguided by an awareness of their global scale consequences. The long-term survival of any global scale project of civilization requires a different mode of planetary-scale behavior where knowledge of planetary-scale impacts feeds back on, and modulates, behavior in an intentional loop, but decisions favoring the sustainability of collectives may not be the same as preferences favored by individuals. A collectives' rankings among a number of choices will not reflect that of individual members.

Planetary behavior dominated by stabilizing feedback between awareness and consequences could represent a new level of planetary intelligence.

Work on the "Astrobiology of the Anthropocene" indicates that technological civilizations engaging in large-scale energy harvesting can trigger strong climate-changing feedbacks. The transition to long-term sustainable forms of these civilizations could have general, generic features involving transitions in planetary intelligence. A characterization of planetary intelligence and its role in planetary evolution might be useful for technosignature studies that could prove useful in astrobiology and SETI.

Consideration of planetary scale cognitive activity goes back to the formative development of biogeochemistry, Earth Systems Science, and Astrobiology. The modern concept of the biosphere can be traced to the work of Vladimir Vernadsky, the founder of both geochemistry and biogeochemistry. Vernadsky saw that the aggregate activity of life on Earth had to be part of a system, the biosphere that couples to the other planetary systems; the atmosphere, hydrosphere, cryosphere, and lithosphere, all driven by the thermodynamics of free-energy gradients.

Vernadsky wrote, "Activated by radiation, the matter of the biosphere collects and redistributes solar energy and converts it ultimately into free energy capable of doing work on Earth. A new character is imparted to the planet by this powerful cosmic force. The radiations that pour upon the Earth cause the biosphere to take on properties unknown to lifeless planetary surfaces, and thus transform the face of the Earth."

For Vernadsky, the biosphere was an emergent phenomenon that appeared with and evolved in tandem with the diversity of individual species. The evolution of such species could only be fully accounted for in the context of the wider biosphere. This emergence always involved some degree of cognitive or cultural activity.

After developing the concept of the biosphere, Vernadsky explored the concept of the Noosphere. Unlike Teilhard de Chardin's theological version of the idea, Vernadsky considered the Noosphere as an emergent shell of influence based on the totality of what he called "cultural biogeochemical energy", and in using the term "culture", he meant *collective* cognitive activity. He held that such activity had always been present in the biosphere from microbes to mammals, but thought the collective cognitive activity in these species, and in the biosphere, was insignificant in both measure and impact until the development of human scientific and industrial activity.

While Vernadsky thought that cultural biogeochemical energy was a minor player in the biosphere, Lynn Margulis an American evolutionary theorist had her own conception of the idea. She believed it played a larger role in planetary evolution via the Gaia Theory she developed with James Lovelock.

THE GAIA HYPOTHESIS

The Gaia Hypothesis as first developed by Lovelock held that Earth's life could maintain global conditions like average temperature within a range that kept the planet habitable. Lovelock thought that this would occur through negative feedbacks between life and planetary geochemistry. These feedbacks would keep perturbations in global conditions in check. What Margulis brought to the collaboration was a focus on the remarkable capacities of microbes to serve as drivers for Gaian feedbacks.

Margulis saw the microbial domains as rich with a kind of pre-intelligence. She wrote, "the view of evolution as chronic bloody competition … dissolves before a new view of continual cooperation, strong interaction, and mutual dependence among life forms."

Gaia was not seen as an organism. Margulis wrote, "Gaia is an emergent property of interaction among organisms, the spherical planet on which they reside, and an energy source, the Sun". This concept of the emergence of a new planetary property from the networked activity of individual players was the central insight of the Gaia Theory. Margulis later wrote "Gaia is the regulated surface of the planet incessantly creating new environments and new organisms…. Less a single live entity than a huge set of interacting ecosystems, the Earth as Gaian regulatory physiology transcends all individual organisms".

The basic principles of Gaia Theory repackaged as "Earth Systems Science" represent the cornerstone of modern approaches to Earth's evolutionary history. What Earth Systems Science took from Gaia Theory was its recognition of the biosphere as a principal driver of planetary evolution, as well as the profound role of collective microbial activity in shaping critical biospheric feedbacks. In this context, life is an emergent property of the physiochemical systems that it is constructed from.

A standard view of emergence says, "the whole is greater than the parts", such that properties and behaviors at collective scales cannot be predicted from or reduced to consideration of the parts alone. While emergence is most often regarded as a property of complex biological and technological systems it is also apparent in physics. Phillip Andersen, a Nobel Laureate in Physics wrote in an essay titled "More is Different" that "The ability to reduce everything to simple fundamental laws does not imply the ability to start from those laws and reconstruct the universe". Emergent properties are not antagonistic to the reductionist view. It is by virtue of the fact that reductionism is possible that we can observe emergent properties at all.

Emergence is often associated with top-down causation where the emergent system creates modes of behavior in its subsystems that would not be possible without new higher-level rules.

The implicate/explicate holographic order represented by the golden mean defines this reductionist concept that gives rise to these emergent properties.

Planetary intelligence would be an emergent, *collective property of the subsystems comprising the biosphere* that induces new modes of behavior on individual parts which implies that life is not a scale-specific phenomenon. It emerges from chemistry and drives the organization of matter from the properties of cells to the planetary scale. The natural boundary for these processes is planetary. The suggestion that intelligence is the mechanism that controls the function, decision-making, and seeming goal-directedness of many living processes is also not scale-specific. It is a general phenomenon that operates at the planetary scale.

The view of life as an emergent phenomenon does not mean that general principles for life cannot be found. The ability to articulate these patterns is important for an effort to use the properties of Earth's biosphere to understand life on other worlds. Information flows appear in living systems from cells to ecosystems to cities, and downward in the form of networks that constrain the behavior and function between system components and subsystems.

Focusing on networks and information flow offers the possibility for developing a more general approach to understanding how behaviors emerge in living systems. Studies of the biochemical networks at three levels of scale, including cells, ecosystems, and the biosphere, reveal a network structure that is common across scales of biological organization, including individuals and communities. These are distinct from random networks implying deeper levels of network structure in living systems. These properties are universal across biochemical networks and depend on size, like the number of compounds that are nodes in the network, and they do not depend on the scale of organization.

The emergence of intelligence operating on the scale of planetary behavior and function is best described by information flowing through the combined technosphere/biosphere geochemical and geophysical networks in different forms. These include processes at higher-scales constraining and determining the behavior of lower-level entities which is what happens in social systems where decisions are dependent on cultural and societal context.

Once a species capable of constructing a technological civilization appears, intelligence by most definitions exists on a planet, but this does not imply the existence of a planetary intelligence as the dominant driver of planetary evolution. Life on Earth emerged almost four billion years ago. By three billion years ago collectives of single-celled organisms existed in large enough quantities to begin affecting the coupled geophysical/geochemical systems. As an example, the formation of methanogens is believed to have changed atmospheric chemistry sufficiently to alter the Earth's radiative properties and trigger the first global glaciation. For the first two billion years of Earth's evolution its atmosphere consisted primarily of N_2 and CO_2 with O_2 acting as a trace gas. It was the evolution of oxygenic photosynthesis by cyanobacteria that led to the atmosphere's Great Oxygenation Event roughly two and a half billion years ago that made O_2 abundant in Earth's biogeochemical networks that allowed far more energetic modes of metabolism.

Microbes also play an essential role in Gaian and Earth Systems Science descriptions of planetary evolution through the establishment of feedback loops that maintain the planet in stable dynamic equilibrium. Known and proposed examples of these feedbacks are climate regulation through biologically enhanced rock weathering; the maintenance of O_2 partial pressures below 30% through methane-producing microbes; climate regulation through cloud-albedo control linked to algal gas emissions; and the biological transfer of selenium from the ocean to the land as dimethyl selenide.

Given the critical role of microbes in establishing these feedback loops, when formulating questions of planetary intelligence the question arises as to whether microbes or their communal networks possess anything like cognition. Do microbes or their collectives "know" anything about the world, rather than just existing in it?

What is meant by knowing and to consider the nature of cognition across all forms of life?

Sara Shettleworth, an American-born, Canadian experimental psychologist and zoologist who focuses on animal cognition sees cognition as the mechanisms by which animals acquire, process, store, and act on information from the environment.

A more extensive definition was given by Pamela Lyon an interdisciplinary visiting research fellow at the Flinders University in Adelaide.

Biological cognition is the complex of sensory and other information-processing mechanisms an organism has for becoming familiar with, valuing and interacting with its environment to meet its existential goals, the most basic of which are survival and reproduction.

There is considerable evidence that bacteria exhibit a range of behaviors associated with cognition in the sense given above. Signal Transduction, the most basic form of sense perception occurs in bacteria in multiple forms, allowing them to sense and respond to a wide array of environmental cues. Bacteria can also communicate through a process known as Auto-Induction where they stimulate changes in their genetic expression when certain environmental molecules reach threshold concentrations, which is the basis of bacterial *quorum sensing*, where advantageous genetic changes in populations are induced at concentrations dependent on population density. Equally important was the discovery of rich social behaviors in species like *Myxococcus xanthus*, "the primate of eubacteria", which is capable of structured, multi-dimensional swarming, pack-like predation, and the use of chemical cues to lure faster-moving prey. Memory and learning, the bedrocks of cognition, have also been present in the bacterial toolkit of behaviors.

There have been forms of cognitive activity on the planet for much longer than there have been animal nervous systems pre-dating the appearance of humans. If the microbes that form planetary feedback loops do collectively know things about their world, then it's possible to ask if this knowing is integrated into higher scale emergent behaviors that represent planetary intelligence.

This concept fits Sheldrake's hypothesis of morphic fields and morphic resonance that leads to a vision of a living developing universe with its own inherent memory. His research on plant development brought him to the conclusion that morphogenesis depends on organizing fields and that biological organization overall depends on fields, variously called biological fields, developmental fields, positional fields, or morphogenetic fields which are not fixed.

They evolve.

In today's biosphere there are semantic information flows that act locally and yield feedback and controls on larger scales like the information encoded in the arrangement of bases in a genome, which is minuscule in size compared to the planet, but has the ability to control metabolic pathways that shape global biogeochemical cycles.

Another example is the mycorrhizal fungi that inhabit the root systems of 80% of land plant species working as mutualistic symbionts.

Given their importance and the ability of these networks to direct nutrients to parts of the forest under stress, the question of self-recognition shows that the root systems of plants belonging to different species, genera, and families, are connected within plant communities creating pathways where semantic information flows across large-scale biomes that are part of an emergent cascade to planetary scales of feedbacks and controls that could prove it capable of producing and maintaining itself by creating its own parts.

The development of agriculture after the last ice age followed by the construction of cities and empires required to support them were the initial stages of building a technological system that eventually reached across the planet. The discovery and application of fossil fuels began an industrial age, which over a few centuries rewired the Earth systems' coupled networks.

Human-induced changes to these systems initiated a new geological epoch called the Anthropocene. While some question if the Anthropocene can be via stratigraphy, substantial evidence exists that Earth has already crossed a boundary where key measures show large-scale human imprints. More than half of the Earth's land surface has been altered for human uses. Current anthropogenic flows of phosphorus are more than a factor of 5 above natural rates.

By the beginning of the twenty-first century, the systems humans constructed were planetary with human beings on every landmass. Our artefacts, from microscopic plastic debris to released fossil CO_2, stretches from the deep ocean to the upper atmosphere, as well as to the Moon and other planetary bodies. Human industrial activity has become a force in the operation of the planetary system and the global impact of human activities requires a level and depth of organization that impacts our definitions of planetary intelligence.

The term "Technosphere" defines the planetary character of human activity and influence, or the activity of an intelligent species that constructs technology. Technosphere is considered "the interlinked set of communication, transportation, bureaucratic, and other systems that act to metabolize fossil fuels and other energy resources." This definition includes flows of material, energy, and information.

Economic activity and organization is often referred to as an "Econosphere" with the use of the suffix "sphere" as an acknowledgement that a given systems' structural and evolutionary characteristics can be considered as planetary phenomenon.

For all its reach, what humans have created with industrial civilizations is unstable. If we consider civilization as a Technosphere consisting of human population and its technological support systems coupled to the other planetary systems including the biosphere, atmosphere, and others, the question of the stability of these coupled systems' forcing and response times become more critical.

A dynamical systems approach was applied to the interaction of an energy-harvesting civilization and the host planet where that energy was harvested. It showed that parameters associated with both the planet's sensitivity to forcing and the civilization's population growth/decline, as a result of energy harvesting, determined the long-term viability of coupled systems.

Population growth drives climate feedbacks producing climate forcings at a faster rate than the planet's own internal mechanisms like weathering, volcanism, and other factors. In this paradigm the activities of the civilization push the planet into new climate states on timescales measured in a few generations, and the global average temperature in these new states are beyond what the technosphere can handle, leading to a rapid reduction in the population within a few generations. If the die-off in the technosphere is large enough in a short enough time measured in generations, then a complex technological civilization may not be capable of maintaining itself.

PHI, EARTH, THE SOLAR SYSTEM, AND BEYOND

Another example of the implicate/explicate holographic order is found in the dimensions of the Earth and Moon which are in Phi relationship, forming a Triangle based on **1.618**.

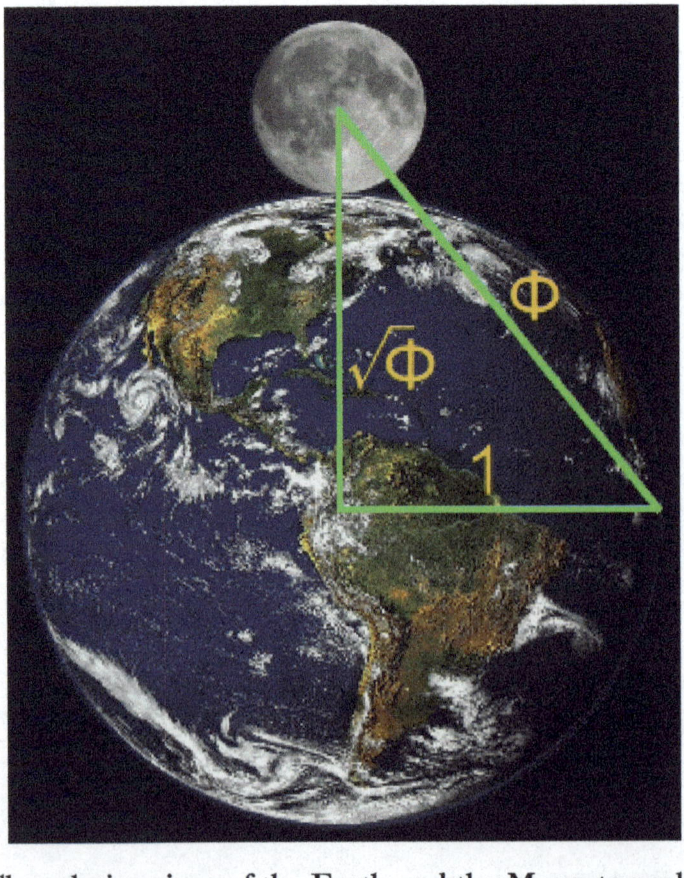

The relative sizes of the Earth and the Moon to scale.

If you Draw a radius of the Earth and draw a line from the center point of the Earth to the center point of the Moon, it will result in the square root of Phi. If you draw a line to connect the two lines it forms a Golden Triangle known as a Kepler triangle. This geometric construction is the same as what appears to have been used in the construction of the Great Pyramid of Egypt.

Pythagoras discovered that a right triangle with sides of length a and b and a hypotenuse of length c has the following relationship:
$a^2 + b^2 = c^2$

A foundational equality of phi has a similar structure:
$1 + \text{Phi} = \text{Phi}^2$
(1+ 1.618… = 2.618…)

By taking the square root of each term in this equality, we have the dimensions of a Kepler Triangle, a right triangle based on this phi equality, where:

Side	Length squared per above	Length, or square root	Length divided by phi so c = 1
a	1	1	1 / Phi
b	Phi	√ Phi	1 / √ Phi
c	Phi^2	Phi	1

This triangle is illustrated below and has an angle of 51.83 degrees or 51 degrees 50 seconds, which has a cosine of 0.618 or phi.

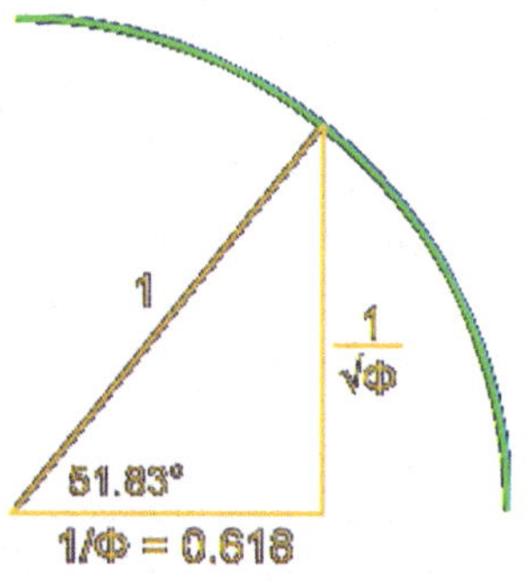

The Pythagorean 3-4-5 triangle is the only right-angle triangle whose sides are in an arithmetic progression. 3 + 1 = 4, and 4 plus 1 = 5. The Kepler triangle is also the only right-angle triangle whose side are in a geometric progression: The square root of phi times **Φ** = 1 and 1 times **Φ** = **Φ**.

There is debate as to the geometry used in the design of the Great Pyramid of Giza in Egypt. Its once flat, smooth outer shell is gone and all that remains is the roughly-shaped inner core, so it is difficult to know with absolute certainty. The outer shell remains at the cone which helps establish the original dimensions.

There is also evidence that the design of the pyramid embodies these foundations of mathematics and geometry:

- Phi, the Golden Ratio that appears throughout nature.
- Pi, the circumference of a circle in relation to its diameter.
- The Pythagorean Theorem – Credited by tradition to mathematician Pythagoras which can be expressed as $a^2 + b^2 = c^2$.

Some planets of our solar system exhibit a relationship to phi as shown by the following table of the time it takes to orbit around the Sun:

	Mercury	Venus	Earth	Jupiter	Saturn
Power of Phi	-3	-1	0	5	7
Decimal Result	0.24	0.62	1.0	11.1	29.0
Actual Period	0.24	0.62	1.0	11.9	29.5

Venus and the Earth are linked in an unusual relationship involving phi. Starting with Mercury representing the basic unit of orbital distance and period in the solar system:

Planet	Distance from the sun in km (000)	Distance where Mercury equals 1	Period where Mercury equals 1
Mercury	57,910	1.0000	1.0000
Venus	108,200	1.8684	2.5490
Earth	149,600	2.5833	4.1521

Ö Period of Venus * Phi = Distance of the Earth
Ö 2.5490 * 1.6180339 = 1.5966 * 1.6180339 = 2.5833

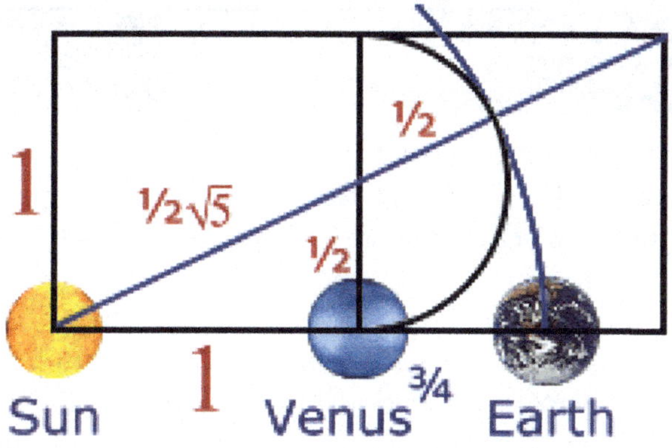

Venus orbits the Sun in 224.695 days while Earth orbits the Sun in 365.242 days creating a ratio of 8/13 which are Fibonacci numbers, or 0.615 which is roughly phi. Five conjunctions of Earth and Venus occur every eight orbits of the Earth around the Sun and every thirteen orbits of Venus.

Mercury orbits the Sun in 87.968 Earth days, creating a conjunction with the Earth every 115.88 days making 365.24/115.88 conjunctions in a year, or twenty-two conjunctions in seven years, which is very close to Pi.

The diameter of Saturn is also close to a phi relationship with the diameter of its rings, and the inner ring division is in a relationship close to phi with the diameter of the rings outside the sphere of the planet. The Cassini division in the rings of Saturn falls at the Golden Section of the width of the lighter outside section of the rings.

A closer look at Saturn's rings reveals a darker inner ring which exhibits the same golden section proportion as the brighter outer ring.

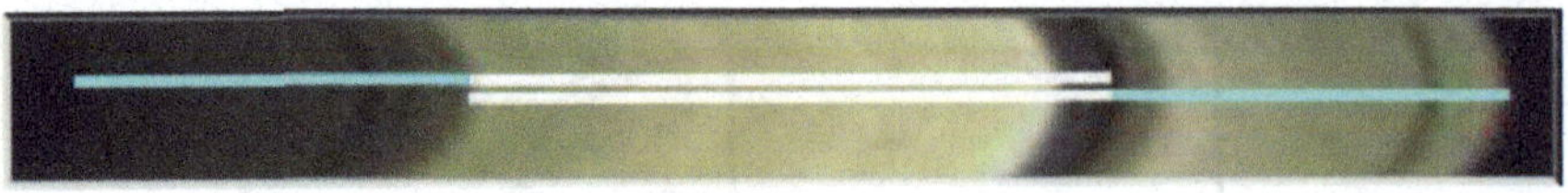

From at least the time of Ptolemy in the 2nd century AD, man believed that the Earth was at the center of the universe until the fifteenth century when Copernicus brought us closer to the truth, moving the Sun to the center of the universe.

Galileo, Kepler and others helped move us from our geocentric view of the universe, but even in the 21st century, science still expresses the planetary relationships of solar system using the "Astronomical Unit," which is the distance of the Earth from the Sun.

Other than the fact that we inhabit the Earth, there is no logical reason to continue basing these geocentric relationships on the third planet of the system.

A new look at the solar system shows the incredible relationships that result when Mercury, the first planet, is used as the base measure for the other planets. Intriguing mathematical relationships appear based on the orbital distances of the planets:

$(\text{Earth}^4 / \text{Venus}^3) + (\text{Venus}^3 / \text{Earth}^4) = 7$

$(\text{Mars}^4 / \text{Earth}^3) + (\text{Earth}^3 / \text{Mars}^4) = 14$

These result from simple geometries that define the distances of planetary orbits:

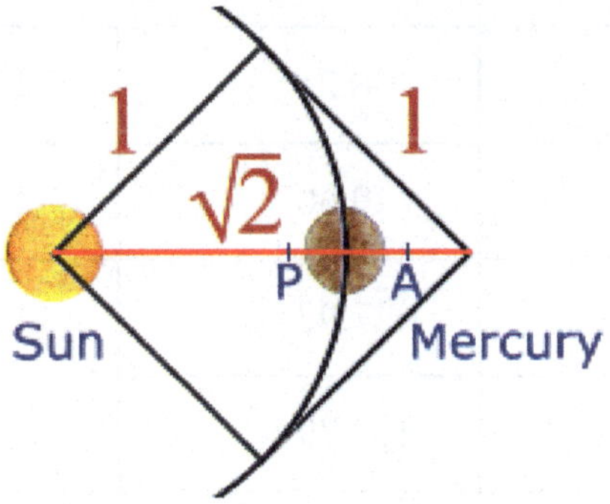

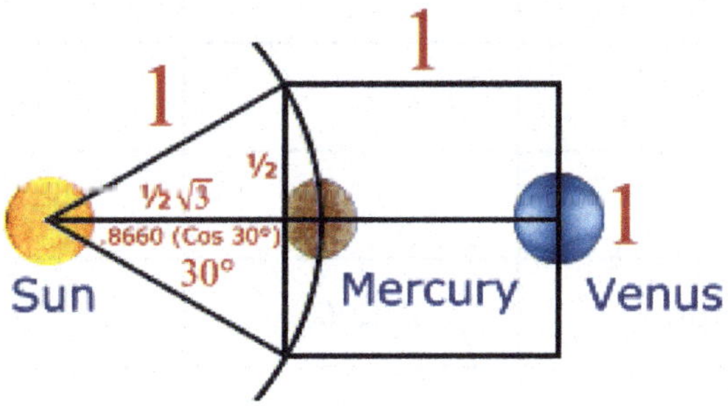

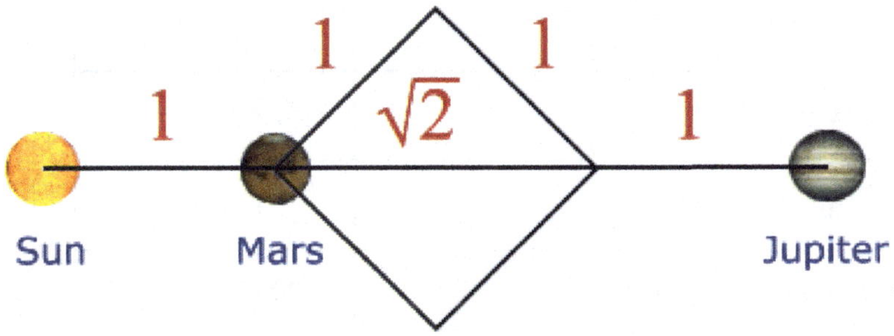

The average of the mean orbital distances of each successive planet in relation to the one before it also approximates phi:

Planet	Mean distance in million kilometers per NASA	Relative mean distance where Mercury=1
Mercury	57.91	1.00000
Venus	108.21	1.86859
Earth	149.60	1.38250
Mars	227.92	1.52353
Ceres	413.79	1.81552
Jupiter	778.57	1.88154
Saturn	1,433.53	1.84123
Uranus	2,872.46	2.00377
Neptune	4,495.06	1.56488
Pluto	5,869.66	1.30580
Total		16.18736
Average		1.61874
Phi		1.61803
Degree of variance		(0.00043)

The standard model of cosmology predicts that the universe is infinite and flat, but cosmologists have suggested that space could be finite and shaped like a dodecahedron, claiming that a universe with the same shape as the twelve-sided polygon can explain measurements of the cosmic microwave background consisting of radiation left over from the big bang that spaces with more mundane shapes cannot.

The cosmic microwave background provides a picture of the universe as it was some 400,000 years after the big bang when the universe cooled down enough for atoms to form, which meant that there were no longer any free electrons to scatter the photons produced in the early universe. Any variations in the temperature of the background radiation reflect variations in the density of the universe at this time.

These temperature fluctuations can be expressed as a sum of spherical harmonics and astrophysicists plot the relative strength of these harmonics as a function of angle. The height and positions of the peaks in this so-called "power spectrum" are related to basic astrophysical properties of the universe.

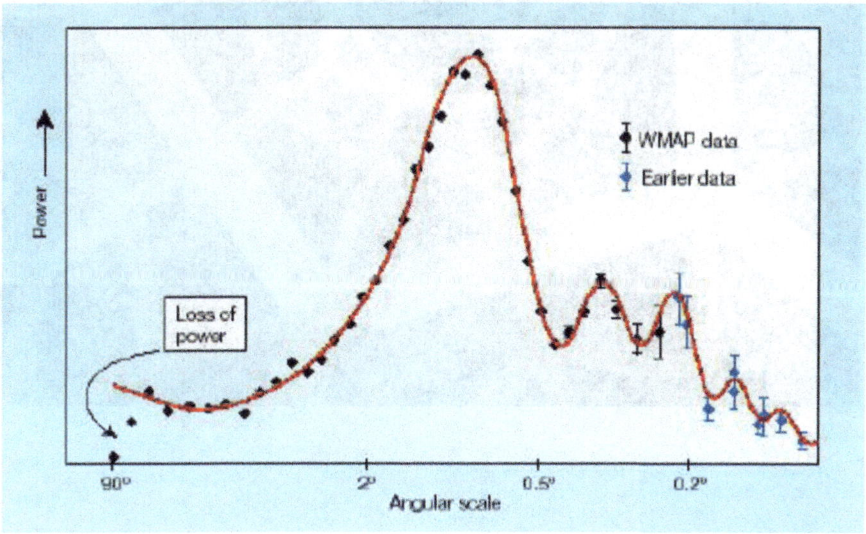

Data from the first year of the Wilkinson Microwave Anisotropy Probe known as the WMAP satellite agreed with the predictions of the standard big bang plus the inflation model of cosmology for regions of space separated by small angles.

Jean-Pierre Luminet of the Observatoire de Paris and colleagues believe that the finite size of the universe itself is responsible for this behavior and show that the predictions of a model in which space consists of 12 curved pentagons joined together in a sphere agrees with the WMAP observations.

A FRACTAL UNIVERSE

A fractal is a mathematical shape whose structures repeat infinitely as you zoom in deeper and deeper. Following this progression, in our Universe many of the structures we see on small scales show up repeatedly on larger scales indicating that we are living in a fractal Universe that follows the implicate/explicate order of the holographic model.

The dark matter halos that form around the largest bound structures we know of are identical to the ones that form around Milky Way-sized galaxies as well as the tiny substructure clumps that exist around smaller galaxies and in intergalactic space itself. On the largest scales in the Universe, gravitation is the only force that matters, and gravitational collapse produces identical structures scaled up or down in size depending on the size of the system.

The idea that if you zoom in far enough you'll encounter a structure that repeats the initial pattern seen on larger scales is mathematically realized in the concept of a fractal. When similar patterns repeatedly emerge at smaller and smaller scales we can analyze them mathematically and see if they have the same statistical characteristics as the larger structures. If they do, it's fractal-like in nature.

Mathematically, most of us are used to real numbers expressed in a decimal format even if that decimal is infinitely long. Even if it never repeats there are more numbers that mathematically exist other than the real ones, like complex numbers that have a real and an imaginary part, which is a real number multiplied by *i*, defined as the square root of -1. This includes the real numbers and takes us beyond the limitations of working with real numbers alone.

The most famous fractal is the Mandelbrot set, which is illustrated in the complex plane, where the x-axis is real and the y-axis is imaginary.

The way the Mandelbrot set works is to consider every possible complex number, *n*, and study the following sequence:

- n,
- $n^2 + n$,
- $(n^2 + n)^2 + n$,
- $((n^2 + n)^2 + n)^2 + n$,

and so on. Each new term is the prior term, squared, plus n. If this sequence doesn't diverge, going to either positive or negative infinity, then the value of *n* is a member of the Mandelbrot set.

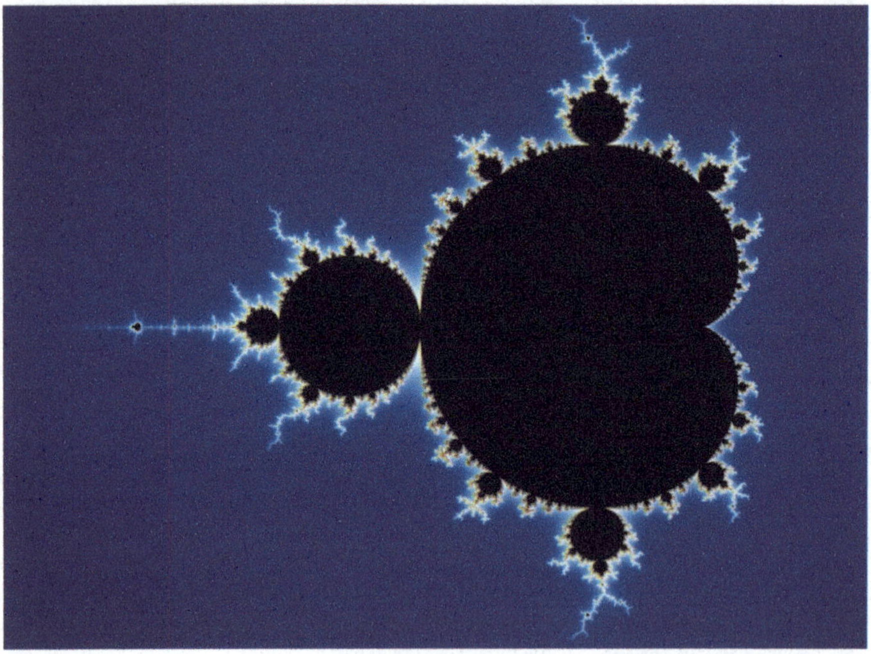

The Mandelbrot set is an example of a mathematical structure with self-similar and quasi-self-similar components to it and is considered the most famous example of a fractal structure.

The way the Mandelbrot set is visualized is by representing the border between what's in the set versus what's outside of it, with color coding showing how far away something is from being a member of the set with brighter colors closer to being in it. This observation shows that many of the emerging patterns are intricate and self-repeating.

When you see a small region with identical properties to the entire set itself, those regions are called "self-similar." If something has almost the same properties as the larger set but with subtle differences, it exhibits quasi-self-similarity and if the small region has identical properties to a larger region, it exhibits true self-similarity.

In the Mandelbrot set you can identify many regions that show both the more common quasi-self-similarity, and true self-similarity which is less common, but still exists. This has been mathematically demonstrated on scales spanning hundreds of orders of magnitude which are larger than the physical scales that take us from the smallest subatomic distances to the entire observable Universe.

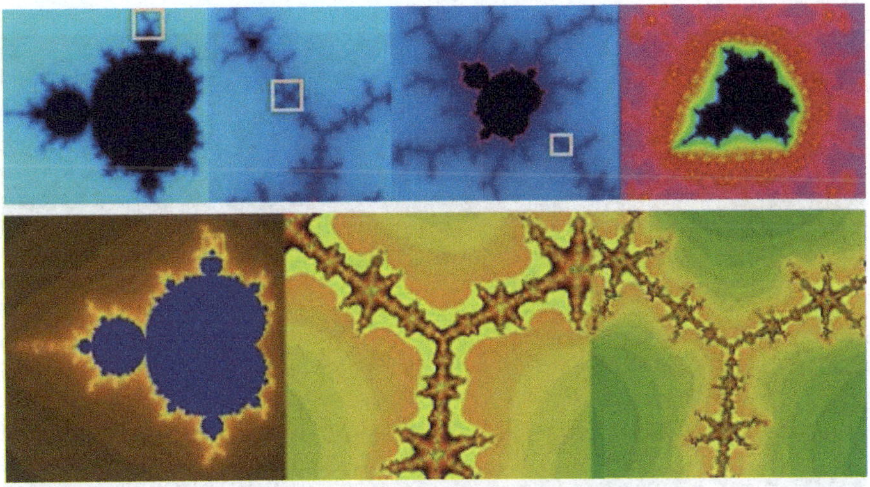

Regions of both quasi-self-similarity (top) and exact self-similarity (bottom) can be found ubiquitously within the Mandelbrot set at a variety of zoom levels. The fact that these mathematical structures repeat was once thought to hold much explanatory promise for our Universe. (Credit: Antonion Miguel de Campos (top); Ishaan Gulrajani (bottom)

From a mathematical perspective the same rules and conditions apply at all scales. Depending on those rules, you might end up with a self-similar structure to the Universe where what appears on large scales also appears on small scales. This perspective highlights two tandem facts about the cosmos.

1. The Universe as a whole appears to have a large amount of unseen invisible mass to it: what we know today as dark matter.

2. The overall spatial curvature of the Universe is consistent with being flat. If you add up all the forms of energy present in it, they equal the critical density, determining the expansion rate.

In physics, astrophysics, and cosmology, we can't simulate the entire Universe with any precision, but we make some simplifying assumptions and simulate the Universe to the best of our abilities under that set of assumptions. One of the more interesting things done was to run simulations of dark matter in the Universe on a variety of scales which all yielded similar results.

According to models and simulations all galaxies should be embedded in dark matter halos whose densities peak at the galactic centers. On timescales of a billion years or more, a single dark matter particle from the outskirts of the halo will complete one orbit. The effects of gas, feedback, star formation, supernovae, and radiation all complicate this environment, making it difficult to extract universal dark matter predictions. (Credit: NASA, ESA, and T. Brown and J. Tumlinson (STScI))

When you start with a Universe uniformly full of dark matter, the same gravitational physics are at play. No matter how uniform you make it there are always tiny imperfections like an atom or molecule that isn't perfectly distributed, a tiny attractive or repulsive force on a subatomic particle, quantum jitter, and other factors. As soon as your system is no longer uniform and perfect uniformity is unstable under the laws of gravity, the overdense regions will attract more matter than the surrounding regions, while the underdense regions give their matter up to the surrounding regions.

If you start with a single overdense clump and allow it to evolve long enough until every particle can complete many orbits of whatever trajectory it's on, you will get a large dark matter halo that is spheroidal and diffuse with its highest density in the center.

The idea of a universal profile for dark matter halos is one of the most exciting predictions in all of self-similarity in cosmology. To be more accurate we need to go beyond a single isolated system and simulate what's happening in a more realistic scenario like dark matter in a Universe that's expanding and filled with a variety of underdensities and overdensities. This is consistent with what we know and observe about the Universe, and if we are making assumptions we should assume something as close to the actual Universe as possible.

Running cosmological simulations will produce a great cosmic web where small scales collapse first as soon as gravity has time to send its influential signal from one overdense region to the surrounding matter. Where larger scales collapse later with smaller scale structure superimposed on top of them, as more time goes on, even larger scales follow suit, giving rise to a completely self-similar Universe.
In this scenario you get mini-halos inside regular halos inside giant halos, all connected by filaments that themselves, given enough time and the right properties will produce their own halos while an even grander web forms on larger scales.

People have been working to measure the fractal dimension of the Universe for decades, attempting to decipher whether it can be well described by one simple fractal parameter, or whether multiple ones are required. The nearby Universe is not a good place to measure this as dark energy has already reared its head for the past six billion years. Looking at objects at a redshift of ~2 or greater, is looking back in time to when dark energy was insignificant, making it the perfect laboratory for studying what type of self-similar properties the Universe had.

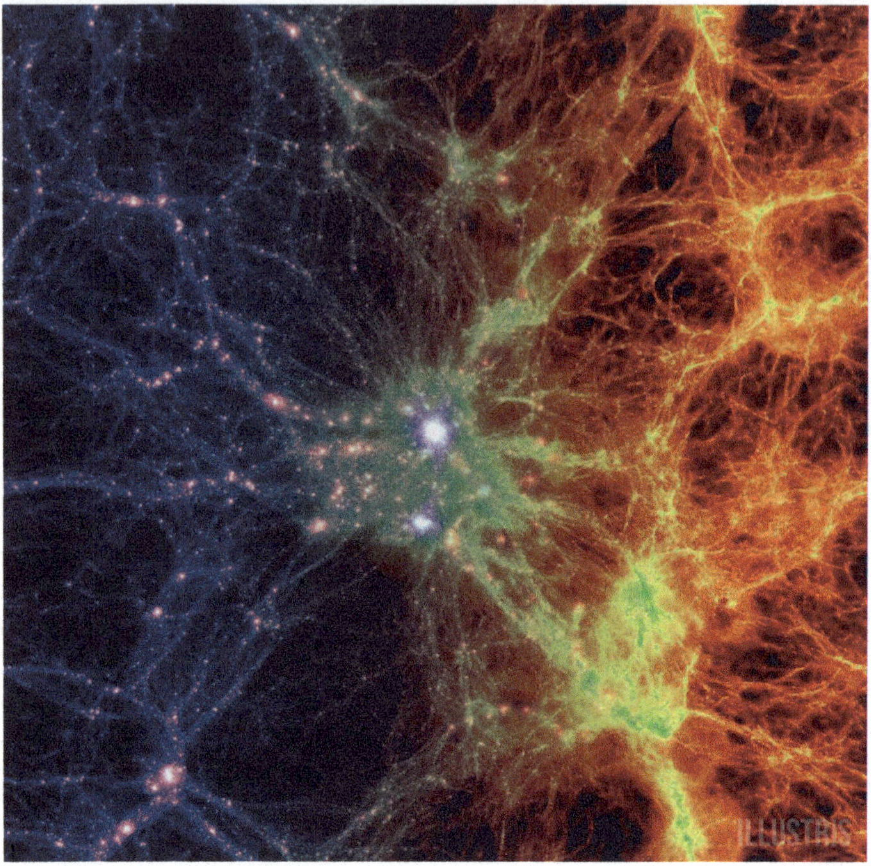

The formation of cosmic structure on both large scales and small scales is dependent on how dark matter and normal matter interact. The distributions of normal matter at left and dark matter at right can affect one another. Things like star formation and feedback can affect normal matter, which exerts gravitational effects on dark matter. (Credit: Illustris Collaboraiton/Illustris Simulation)

COSMOMETRY

Cosmometry is an emerging field of knowledge and application that synthesizes a millennia of observation, study, theory, and direct experience, both scientifically and metaphysically, of the fundamental patterns, structures, processes and principles of universal creation. Whether through physics, music, bio-energetics, visionary art, or shamanism, a unified understanding of cosmic coherence is arising in consciousness at new levels of integration and wholeness.

The premise of cosmometry is that the universe is a fractal-holographic phenomenon composed of a synergetic interplay of energy and consciousness that coexists in the ongoing creation of physical and metaphysical existence at all scales which is also the fundamental insight of many spiritual and mystical traditions. Science and Spirit are converging into a unified awareness where the purely metaphysical and the purely physical pursuits of knowledge become seamlessly integrated.

This fractal-holographic world view endorses the implicate/explicate order stating that "the same patterns are repeated at all scales", which is fractal, and "the whole is present everywhere at all times", which is holographic. Cosmometry includes a synergetic world view which states that an innate attribute of the cosmos is "the behavior of whole systems that is greater than and unpredictable from the behavior and sum of the constituent parts when viewed separately."

At its foundation cosmometry is a study of the fundamental patterning of cosmic manifestation, with cosmic meaning all phenomena at all scales, micro to macro being composed of three energetic-synergetic components.

Vector Tensegrity – geometric pattern and structure like that seen in crystalline forms.

Torus Flow Process – primary form of energy in motion as a discrete entity like that seen in whirlpools.

Field Patterning – standing-wave interaction of vector and torus dynamics like that seen in patterns in water.

Although these three components are unique patterns and forms in and of themselves, they are ultimately a single phenomenon in the unity of wholeness, and the most basic model of cosmometry is this unity of wholeness.

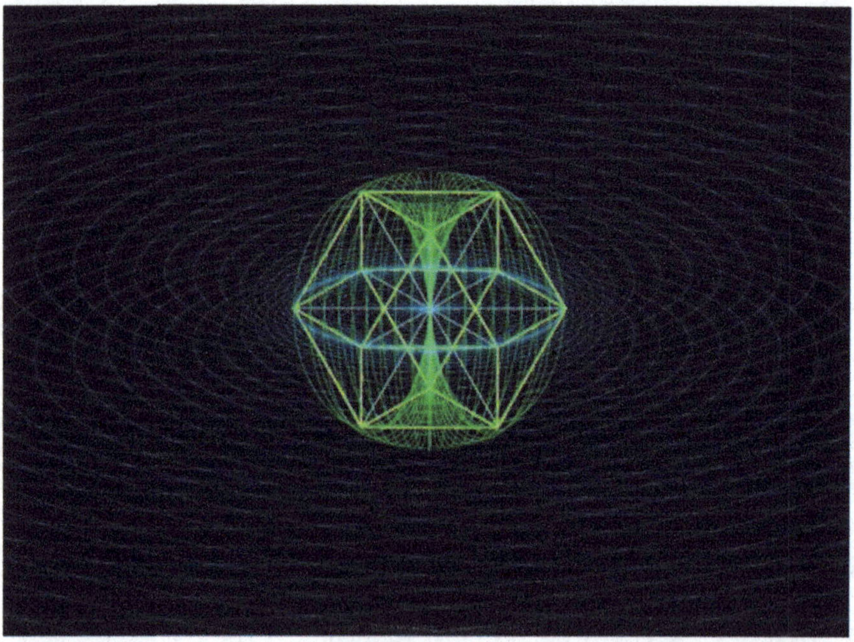

A rendering of cosmometry in an idealized balanced expression

Foundational to the cosmic phenomena observed in photonic and atomic radiation, conceptually determined to exist as in dark matter/energy, and in direct experience as in consciousness itself is a unified field of infinite energy and creative potential. Commonly known as the zero-point field or vacuum of space called plenum, meaning "fullness", it is the source state that precedes manifest reality in physical and metaphysical conceptions. This unified field has a specific cosmometry or geometry where all energetic radiation, gravitation, magnetism, tension, pressure, thought, and feeling resolves into a state of perfect and absolute equilibrium of zero differential and fluctuation. The name of this cosmometric pattern is the Isotropic Vector Matrix, or IVM. Isotropic means "all the same". Vector means "line of energy". Matrix means "pattern of energy lines". The IVM is a pattern of energy lines that are the same everywhere.

When energy comes into a state of perfect and absolute equilibrium, all differentiation becomes zeroed out, resulting in a Unified Wholeness that is beyond the concept of unity. Buckminster Fuller's synergetics state that *unity is plural and at minimum two*, a tenet which is embraced in cosmometry. The unified field, or The "Oneness" of spiritual awareness is the Tao that cannot be named or described because to do so collapses it into a finite "something".

The pioneering work of Buckminster Fuller who first discerned this geometry in 1917 called it the Vector Equilibrium referred to as VE in 1940 in an effort to understand this concept of perfect equilibrium. As a geometric form it is known as a cuboctahedron, which is a polyhedron with 8 triangular faces and 6 square faces. A cuboctahedron has 12 identical vertices, with 2 triangles and 2 squares meeting at each, and 24 identical edges, each separating a triangle from a square. As an expression of ultimate equilibrium Fuller named it a cuboctahedron to describe its energetic condition of vectors that are all the same in length to the center and around the circumference. The VE is the only geometry that satisfies this condition. Others like the platonic forms, have equal-length edges, but shorter or longer vectors to their center point.

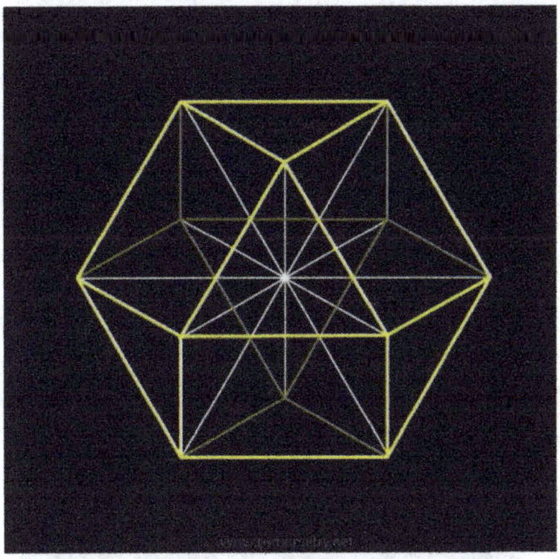

The Vector Equilibrium (VE) or cuboctahedron

Due to having square faces, with squares being unstable

structurally, the VE has the inherent ability to "collapse" in a spiraling motion in what Fuller called "jitterbugging" because it can rapidly collapse and expand in both left and right spirals, pulsating and jitterbugging like a dancer from the thirties. As it does so it transforms through phases that include symmetrical articulation of the icosahedron, dodecahedron, octahedron, cube, and tetrahedron.

In other words, the VE inherently has the potential to create all of the platonic solids in one swift motion.

The VE is the most basic symmetry of the Isotropic Vector Matrix, and is essentially the focal point of all manifest energy events, whether protons, electrons, crystals, planets, stars, or galaxies. Every one of these events is the result of an impulse that introduces tension/pressure dynamics or differentials that collapse the zero-point IVM/VE cosmometry into a discrete phenomenon with its own unique focal point that is the center around which it exists. This center point is the "singularity" that connects the entity to the unified field in its zero-phase. Everything is always connected to the unified field in its zero-phase, and in this way, everything is always connected to the center of everything else, making it implicate/explicate which are defining holographic qualities.

When energy becomes manifest by collapsing out of IVM zero-point equilibrium, it sets up a polarized dynamic, creating the fundamental form of energetic flow called a **Torus**. Everything, from atomic "particles" to electromagnetic fields, weather patterns, trees, humans, ecosystems, planets, stars, and galaxies, are all toroidal in nature. The Universe is a seamlessly embedded fractalization of torus energy flow.

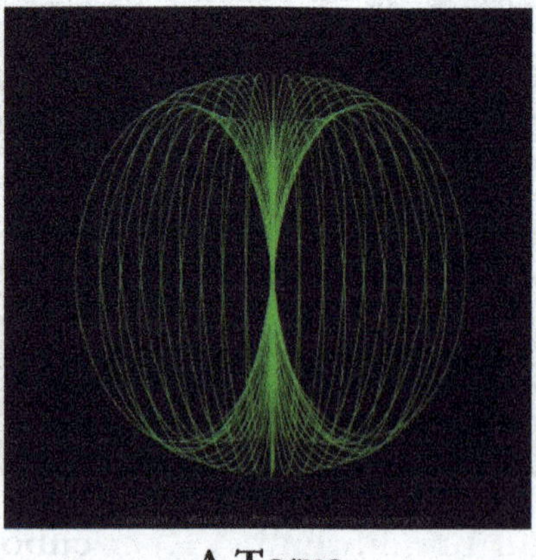

A Torus

Arthur Young, the inventor of the Bell helicopter, was one of the first scientists to study the torus as a fundamental energetic form of the Universe. He described the simple truth of the torus as a model of universal wholeness, stating, *"The self in a toroidal Universe can be both separate and connected with everything else."*

The characteristics of the torus are of primary importance to understanding how healthy and sustainable living systems are formed at all scales of existence.

- It is made from the medium in which it exists
- It is self-generating and self-sustaining
- It exchanges information and resources throughout itself and with the environment in which it exists
- It has a continuous feedback loop and is self-reflexive
- It is both an autonomous entity and integral part of a larger wholeness
- It is centered by a singularity that connects it with the entire Universe

These and other characteristics are what evolution biologist Elisabet Sahtouris observed in nature as Features of Healthy Living Systems. Understanding and applying these features to the design of our human-contrived systems of economics, governance, health, education, and energy is paramount at this time as we witness the demise of these systems due to the lack of these features in their current designs and structures. If we can accomplish this it will bring balance, wholeness, and harmonization with the universal flow of energy, resulting in the stabilization and natural sustainability of humanity and all life on Earth and beyond.

The third primary component of cosmometry is the patterning of energy as it manifests in and around all entities. This Field Patterning seamlessly scales from micro to macro, and quantum to galactic with maximum efficiency and harmony. By exploring this patterning we can observe one of the most basic aspects of the fractal-holographic nature of the cosmos; the ubiquitous presence of the PHI ratio. It appears that the general manifest field of spacetime is based on the phi ratio in its cosmometry.

In the basic model of cosmometry this field patterning is called a Double Spiral which is common in nature in the Fibonacci sequence and integral to toroidal dynamics. It can be seen as the cross-section of a torus that also maps a variety of energy flow patterns.

A Double Spiral seen from an oblique angle

The next phenomenon that emerges is called the Phi Boundary Condition which is the distinct circle that can be seen in the following illustration as well as the one above.

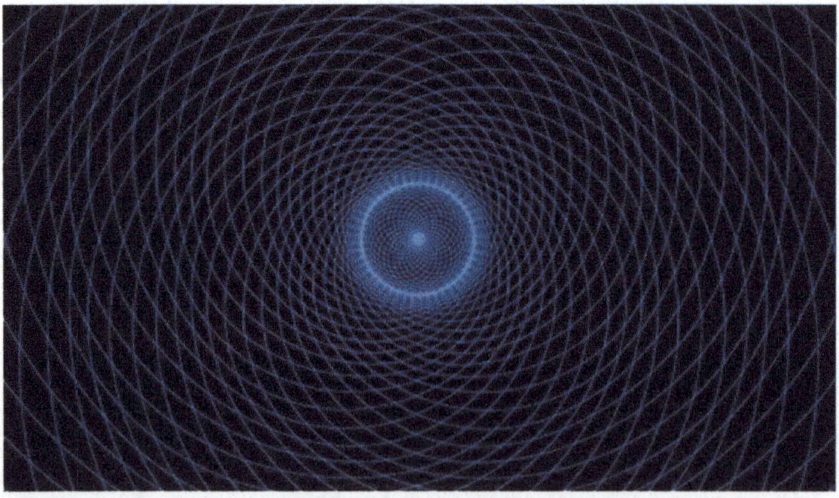

The circle is a synergetic result of spinning the phi spiral, creating a boundary condition

This synergetic phenomenon creates an energetic boundary that defines the interface between the inside and outside of a system or entity from a purely energetic perspective. In order for anything to come into existence in form there naturally manifests a boundary defining itself and its environment which is what the Phi Boundary Condition fundamentally represents.

Although in nature this boundary condition is often distorted by the interaction of a number of dynamic forces, in the representation shown above, a perfect circle is present which can symmetrically align with both the four hexagonal planes of the VE creating a spherical VE or what's called a Genesa Crystal, as well as the "surface" membrane of the torus.

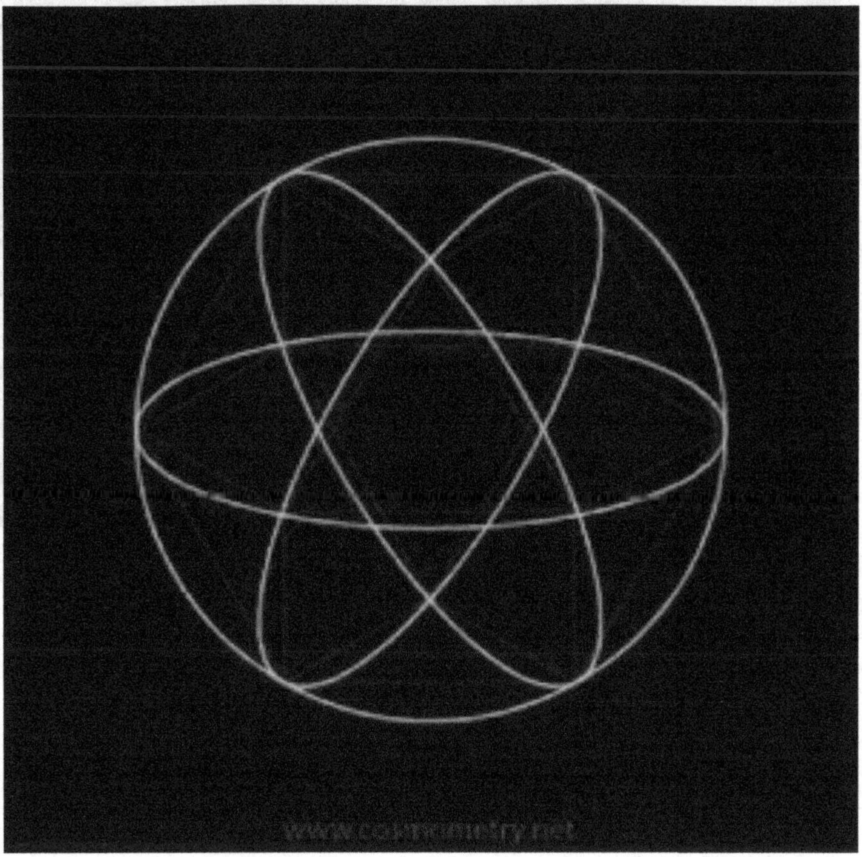

Spherical VE / Genesa Crystal

The combination illustrated below shows the basic unified model of cosmometry. The VE structural tensegrity, Torus flow form, and Phi Double Spiral field patterning depict the foundational integrity of Spacetime and Unified Field energy in its most idealized expression. This model represents the fundamental wholeness that fractally embeds expansively and contractively, and is holographically present everywhere at all scales.

Unified Model of 4 Phi Double Spirals, VE and Torus

UNIFYING THEORIES OF CONSCIOUSNESS

We know we are conscious because we experience and feel things, yet scientists and great thinkers are at a loss to explain what consciousness is and equally baffled about where it comes from.

"Consciousness — or better, conscious experience — is obviously a part of reality," said Johannes Kleiner, a mathematician and theoretical physicist at the Munich Center For Mathematical Philosophy, Germany. "We're all having it but without understanding how it relates to the known physics, our understanding of the universe is incomplete."

With that in mind, Kleiner hopes that math will enable him to precisely define consciousness. Working with Sean Tull, a mathematician at the University of Oxford, the pair are driven by a philosophical point of view called panpsychism; the view that the mind or a mindlike aspect is a fundamental, ubiquitous feature of reality. It is also described as a theory of the mind as a fundamental feature of the world that exists throughout the universe and is inherent in even the tiniest pieces of matter, an idea that suggests the fundamental building blocks of reality have conscious experience.

Integrated Information Theory, or IIT is an attempt to define consciousness with an equation originally put forth by Italian neuroscientist Giulio Tononi in 2004. IIT is designed to measure consciousness in anything, including computers, rocks, and extraterrestrials as easily as in brains. By making a universal claim of what consciousness is, it differs from more biology centered theories.

IIT can free itself from the specific physical features of the brain because its inspiration comes from a different source of introspection. By reflecting on the first-person conscious experience, Tononi came up with five important traits fundamental to consciousness which are the "axioms" that IIT is built on.

The first axiom is the fact that consciousness exists. Others include the observation that a conscious experience is composed of multiple distinct sensations. The experience is specific and appears to us as an integrated whole and it is uniquely what it is – no more or less.

Tononi considered what kinds of information-processing systems could give rise to these axioms of experience, and through this was able to map the axioms to mathematical terms. The end result is a unified measure of what is called "integrated information", a value Tononi symbolized with the Greek letter phi.

Aside from representing the golden ratio, phi also represents numerous other qualities and quantities in physics, mathematics, and other sciences. In this case, total phi indicates just how intermixed the information in a system is. The right kind of intermixing is supposed to give rise to the richness and wholeness of experience. According to IIT, the higher the phi that a system has, the more conscious it is.

Calculating the phi for a system with any reasonable amount of complexity is nearly impossible. For the human brain it would require conducting an endless amount of experiments to probe how the different sub-structures of the brain interact. Even if that could be done, a long and grueling series of computations would be needed. To overcome this hurdle, multiple approximations to phi have been devised, and through this it is possible to make educated guesses about the phi in a system. This has been used to explain why certain brain states lead to more conscious experience than others. As an example, during sleep, the ability of neurons to communicate effectively is interrupted which makes the brain less able to integrate information, resulting in lower phi.

According to Tononi's theory, similar reasoning can explain the unconsciousness that comes with seizures as well. The theory also makes some surprising, predictions. As an example, the phi of an average thermostat is small, but still not zero which implies that the device regulating room temperature has some amount of conscious experience. Some simple devices, if built right, can have a value of phi higher than the estimated phi of the human brain. These counterintuitive conclusions make some scientists and philosophers skeptical of IIT.

Another critique of the theory is targeted at its axiomatic basis. According to this argument, the axioms Tononi chose aren't the only ones that a theory of consciousness can be built on, and his way of mapping these axioms to mathematics isn't the only, or best way. The problem is that if the foundations of IIT are arbitrary, how can we trust the conclusions that come from them, especially when they surprise us?

An informal survey of consciousness scientists revealed that IIT was not the favored theory among experts. It came in fourth after two other theories and the catch-all category of "other". The same survey found that IIT fared better among nonexperts, and was rated first among the subset of non-experts who felt they had enough knowledge to respond. From the outside, the theory looks well founded because of the hard math behind it, but consciousness is a hard concept to tame, which makes IIT's effort to submit it to orderly scientific inquiry a step in the right direction.

If researchers can answer how our brains give rise to subjective experience, there's a chance their mathematical model could also extend to inanimate matter. Oxford physicist Sir Roger Penrose was among the first academics to propose we go beyond neuroscience when looking at consciousness into panpsychism. Penrose says we should consider the role of quantum mechanics and view human consciousness as non-algorithmic and a product of quantum effects. This idea evolved in collaboration with anesthesiologist and psychologist Stuart Hameroff into a hypothesis called Orchestrated Objective Reduction, or Orch OR that claims consciousness is likely due to quantum vibrations in microtubules deep within brain neurons as opposed to the view that it is due to connections between neurons. Orch OR suggests that there is a connection between the brain's biomolecular processes and the basic structure of the universe.

IIT backs panpsychism because even a proton can possess phi according to the theory, and just as an apple, thermostat, and computer can possess it, so can your chair, desk, and all manner of other things across the universe.

"When it comes to experimental evidence, there are several independent studies which point at a correlation between integrated information and consciousness," Kleiner said. "The theory consists of a very complicated algorithm that, when applied to a detailed mathematical description of a physical system provides information about whether the system is conscious or not, and what it is conscious of.

The mathematics is such that if something is conscious according to the theory, then the components which make up that system can't have conscious experiences on their own. Only the whole has conscious experience, not the parts. Applied to your brain, it means that some of your cortex might be conscious but the particles that make up the cortex are not themselves conscious.

If there is an isolated pair of particles floating around somewhere in space, they will have some rudimentary form of consciousness if they interact in the correct way."

According to IIT the universe is full of consciousness, but does it have implications for the physical part of the universe? The math of the theory says it does not. A physical system will operate independently whether it has a conscious experience or not.

"This is at odds with the metaphysical underpinning of the theory which is strongly idealist in nature," Kleiner said. "It puts consciousness first and the physical second. We might see some change in the mathematics at some point to take this underpinning more properly into account.

They would reject the idea that consciousness is separate from or more primary than the physical and they would say consciousness is nothing but a specific physical phenomenon which emerges from the interaction of the fundamental physical quantities in certain conditions."

The math version of IIT is intended to be called a fundamental theory of consciousness. "It tries to weave consciousness into the fundamental fabric of reality, albeit in a very specific way," said Kleiner, "yet if it is proven that consciousness plays a causal role in the universe, it would have huge consequences for the scientific view of the world. It could lead to a scientific revolution on a par with the one initiated by Galileo Galilei."

COULD THE UNIVERSE BE A NEURAL NETWORK?

For years physicists have attempted to reconcile quantum mechanics and general relativity. The first posits that time is universal and absolute, while the latter argues that time is relative, linked to the fabric of space-time. A physics professor at the University of Minnesota named Vitaly Vanchurin believes that artificial neural networks can exhibit approximate behaviors of both universal theories. Since quantum mechanics "is a remarkably successful paradigm for modeling physical phenomena on a wide range of scales," he writes, "it is widely believed that on the most fundamental level the entire universe is governed by the rules of quantum mechanics and even gravity should somehow emerge from it.

We are not just saying that the artificial neural networks can be useful for analyzing physical systems or for discovering physical laws, we are saying that this is how the world around us actually works," Vanchurin says. "With this respect it could be considered as a proposal for the theory of everything, and as such it should be easy to prove it wrong."

Vanchurin says that the universe is a neural network in two ways. The first way is to start with a precise model of neural networks and study its behavior within the limit of a large number of neurons. In this case the equations of quantum mechanics describe the behavior of the system near equilibrium, and equations of classical mechanics describe how the system behaves further away from equilibrium. As far as we know quantum and classical mechanics are exactly how the physical world works.

The second way is to start from physics. Quantum mechanics works well on small scales and general relativity works well on large scales, but it has not been possible to reconcile the two theories in a unified framework. This is known as the problem of quantum gravity.

Compounding matters is the problem of knowing how to handle observers which is known as the measurement problem in the context of quantum mechanics, and the measure problem in the context of cosmology.

Scientists debate that there are not two, but three phenomena that need to be unified: quantum mechanics, general relativity, and observers. Nearly all physicists say that quantum mechanics should be the main field of study and everything else should emerge from it, but nobody knows how to do that. Another possibility is that a microscopic neural network is the fundamental structure, and everything else, like quantum mechanics, general relativity, and macroscopic observers, emerges from it.

In other words, the implicate/explicate order inherent in holograms and the mathematical perfection of the Divine Proportion provide a viable model for this supposition.

Vanchurin's theory says that, "everything you see around you is a neural network and so to prove it wrong all that is needed is to find a phenomenon which cannot be modeled with a neural network. But if you think about it, it is a very difficult task mainly because we know so little about how the neural networks behave and how the machine learning actually works. That was why I tried to develop a theory of machine learning in the first place."

A team of theoretical physicists working with Microsoft published a paper that describes the universe as a self-learning system of evolutionary laws, essentially stating that we live inside a computer that learns. Dubbed "The Autodidactic Universe," the paper argues that the laws governing the universe are an evolutionary learning system; in essence a computer, and rather than exist in a solid state, it perpetuates through a series of laws that change over time.

The researchers explain the universe as a learning system by invoking machine learning systems. In the same way that we can teach machines to perform unfolding functions over time in order to learn, the laws of the universe are essentially algorithms that work in the form of learning operations.

As an example, the researchers say, "When we see structures that resemble deep learning architectures emerge in simple autodidactic systems might we imagine that the operative matrix architecture in which our universe evolves laws, itself evolved from an autodidactic system that arose from the most minimal possible starting conditions?" We understand the laws of physics as we observe them, so it makes sense that the original physical law would be simple, self-perpetuating, and capable of learning and evolving.

Maybe the universe didn't begin with a Big Bang, but with a simple interaction between particles. The researchers allude to this humble origin by stating "information architectures typically amplify the causal powers of rather small collections of particles."

These scientists describe the ever-evolving laws of the universe as irreversible. "One implication is that if the evolution of laws is real, it is likely to be unidirectional, for otherwise it would be common for laws to revert to previous states, perhaps even more likely than for them to find a new state. This is because a new state is not random but rather must meet certain constraints, while the immediate past state has already met constraints. A reversible but evolving system would randomly explore its immediate past frequently. When we see an evolving system that displays periods of stability, it probably evolves unidirectionally."

In illustrating these points, the researchers invoke the image of a forensics expert attempting to recreate how a given program came to a result. In one example, the expert could check the magnetic marks left on the hard disk. In that way, the results of the program are reversible as a history of their execution exists, but if the same expert tried to determine the results of a program by examining the CPU, it would be much more difficult to do as there is no intentional internal record of the operations a CPU runs.

In this research cosmologists explained the history of the universe as one of self-teaching, autodidactic algorithms. The scientists, including physicists from Brown University and the Flatiron Institute say the universe has probed all the possible physical laws before landing on the ones we observe around us today.

The scientists say, "The notion of 'learning' as we use it is more than moment-to-moment, brute adaptation. It is a cumulative process that can be thought of as theorizing, modeling, and predicting. For instance, the DNA/RNA/protein system on Earth must have arisen from an adaptive process, and yet it foresees a space of organisms much larger than could be called upon in any given moment of adaptation."

In an analogy to the research of Charles Darwin who studied all the different ways animals specialized in order to thrive in different environments, questions arise like why we have one monolithic body of laws of physics rather than a bunch of specialized kinds of finches? This is an old question that dates back to at least 1893, when a philosopher first posited "natural selection," but this one applies to the laws of the universe.

In the paper, scientists define a slew of terms including how they define "learning" in the context of the universe around us. The universe is made of systems that each have processes to fulfill every day.

Each system is surrounded by an environment made of different systems. Imagine standing in a crowd of people where your immediate environment is made of other people, and each of *their* environments is made of you and other things.

Evolution is a kind of learning, so when we suggest the universe has used natural selection as part of the realization of physics, we invoke that specific kind of learning, which raises the question; does something have to have consciousness in order to learn? Learning needs to be carefully defined in order to make that the case. Organisms and systems constantly show learning outcomes like more success or higher rates of reproduction.

According to the researchers, "In one sense, learning is nothing special; it is a causal process, conveyed by physical interactions, yet we need to consider learning as special to explain events that transpire because of learning."

Christof Koch, a leading researcher on consciousness and the human brain famously called the brain "the most complex object in the known universe." With a hundred billion neurons and a hundred trillion connections, the brain is an intricately complex object that lives up to this description, but there are also plenty of other complicated objects in the universe.

Galaxies can group into enormous structures called clusters, superclusters, and filaments that stretch for hundreds of millions of light-years. The boundary between these structures and neighboring stretches of empty space called cosmic voids are extremely complex. Gravity accelerates matter at these boundaries to speeds of thousands of kilometers per second, creating shock waves and turbulence in intergalactic gases. These facts indicate that the void-filament boundary is one of the most complex volumes of the universe as measured by the number of bits of information it takes to describe it.

Astrophysicists and neuroscientists compared the complexity of galaxy networks and neuronal networks and discovered that not only are the complexities of the brain and the cosmic web similar, so are their structures.

The task of comparing brains and clusters of galaxies requires dealing with data obtained in drastically different ways: telescopes and numerical simulations on the one hand, electron microscopy, immunohistochemistry, and functional magnetic resonance on the other. It also requires analyzing enormously different scales. The entirety of the cosmic web which is the large-scale structure traced out by all of the universe's galaxies extends over tens of billions of light-years which is 27 orders of magnitude larger than the human brain, and one of these galaxies is home to billions of actual brains. If the cosmic web is at least as complex as any of its constituent parts, we might conclude that it has to be at least as complex as the brain.

The researchers used 4 micrometer thick slices of the human cortex which is the outer layer of the brain responsible for processing language, sensory information, thought, memory, and consciousness. These were compared to 25 megaparsec (1 parsec = approx. 3.26 light years) thick "slices" of Universe taken from a computer simulated volume of 1 million cubic megaparsecs of space. The slices of brain and Universe are relatively comparable in thickness given that both are 27 orders of magnitude in size different from each other.

The slices were examined depending on the scale and the similarity in structure wasn't always apparent, but at 40x magnification in the brain tissue, the researchers saw similarities in structure. 40x magnification represents a scale of 0.01-1.6mm in the brain and 1-100 megaparsecs in the Universe. Here the neuronal network appears like the galaxy clusters and the similarity of the networks can be objectively measured and compared using two techniques.

The first is "network degree centrality" which measures the lengths of the network connections and degree of connectivity in a given network. The nucleus or center of a neuron is smaller in radius than the length of connecting axions and dendrites. Similarly, galaxy clusters are smaller in radius than the length of connecting filaments.

The second method to compare both networks is the "clustering coefficient" which quantifies the amount of structure adjacent to each connection node whether neuron or galaxy cluster, and compares that structure to a random point within the network. This comparison contrasts organization vs randomness in both networks.

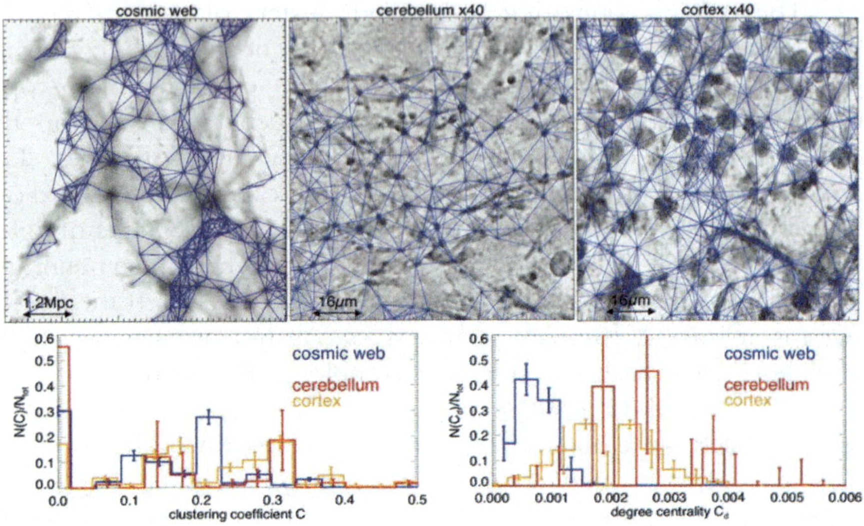

Zoomed details of the reconstructed connections between nodes for three example of networks with lines superimposed to the density contrast maps. Bottom panels show distributions of clustering coefficient and of degree centrality for all slices.

The total number of neurons in the human brain falls within the same parameter of the number of galaxies in the observable universe and the concept of emergence makes the comparison possible. Many natural phenomena are not equally complex at all scales. The expansive network of the cosmic web becomes evident only when the sky is surveyed over its largest extent. On smaller scales, with matter locked into stars, planets, and possible dark matter clouds, this structure is lost. An evolving galaxy does not care about the dance of electron orbitals within atoms, and electrons move around their nuclei without regard to the galactic system they reside in. In this way, the universe contains many systems nested into systems, with little to no interaction across different scales, which allows us to study physical phenomena as they emerge at their own natural scales. This system nestled within systems follows the implicate/explicate order of a hologram.

The building blocks of the cosmic web are the self-gravitating halos of stars, gas, and theoretical dark matter. In total, the number of galaxies in the observable universe should be on the order of 100 billion. The balance between the accelerating expansion of the fabric of spacetime and the pull of self-gravity gives this network its spider-web-like pattern. Ordinary and dark matter condense into string-like filaments, and clusters of galaxies form at filament intersections, leaving most of the remaining volume empty, resulting in a structure that looks biological.

Cortical gray matter representing over 80 percent of brain mass contains about 6 billion neurons which comprise 19 percent of brain neurons and nearly 9 billion non-neuronal cells. The cerebellum has about 69 billion neurons which is 80.2 percent of brain neurons as well as roughly 16 billion non-neuronal cells. The total number of neurons in the human brain falls in the same ratio as the number of galaxies in the observable universe.

The distribution of fluctuations in the cerebellum at 0.1-1 mm scales is reminiscent of the galaxy distribution on hundreds of billions of light-years. At the smallest scales available to microscopic observation around 10 μm, the morphology of the cortex more closely matches the one of galaxies on scales of a few hundreds of thousands of light-years.

This implicate/explicate similarity in memory capacity means that the body of information that comprises their entire life experience is stored in a human brain and encoded into the distribution of galaxies in our universe.

The cosmic web is more similar to the human brain than it is to the interior of a galaxy, and the neuronal network is more similar to the cosmic web than it is to the interior of a neuronal body. Despite differences in substrate, physical mechanisms, and size, the human neuronal network and the cosmic web of galaxies are holographically similar.

This high degree of similarity suggests that the self-organization of both complex systems are being shaped by similar principles of network dynamics, despite their radically different scales and processes. Both neurons and galaxies have nodes connected by filaments. By studying the average number of connections in each node and the clustering of connections in nodes, researchers concluded that there were definite "agreement levels" in connectivity, suggesting the two networks grew as a result of similar physical principles.

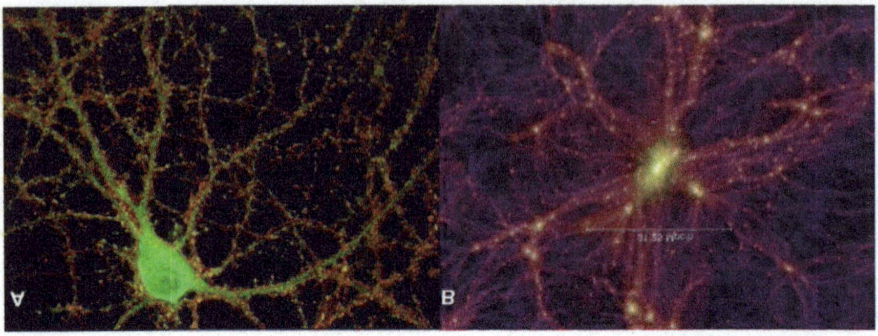

Hippocampal mouse neuron with synaptic connections on the left with a central cell body approximately 10μm in diameter. The cosmic web on the right scaled at 1.4 × 1024 m.

Neurons are networked, communicating with each other through connections called axons and dendrites. There are on the order of 100 trillion connections between neurons forming the neural network that creates who we are.

The Universe is networked as well. The Universe we see with our scientific equipment is referred to as the "Observable Universe" approximately 90 billion light years in diameter and containing on the order of hundreds of billions to a few trillion galaxies. These galaxies, like the Milky Way, are collections of billions of stars, which are themselves grouped into galaxy clusters. The Milky Way is part of the "Local Group" which contains the neighboring Andromeda and Triangulum galaxies as well as 50 other galaxies which are in turn part of a larger group called the Virgo Supercluster. The space between groups and clusters is not empty. It hosts connecting filaments of both ordinary and dark matter that stretch for millions of light years. In this way, the Universe can be thought of as a giant network of galaxy clusters all interconnected similarly to neural networks in the brain in the Cosmic Web.

NGC 6888 the "Crescent Nebula" resembles a giant brain in space.

INFINITE REFLECTIONS FROM SOURCE TO THE HOLOGRAPHIC HEART

A hologram's point of origin comes from a highly focused source beam of light generated by optical amplification based on the stimulated emission of electromagnetic radiation known as laser.

This amplified laser beam is split into two separate beams. The first beam is bounced off the object to be photographed, acting as a reference beam from the source. The second beam passes through a diffusing lens before colliding with the first beam resulting in the interference pattern that is recorded on film.

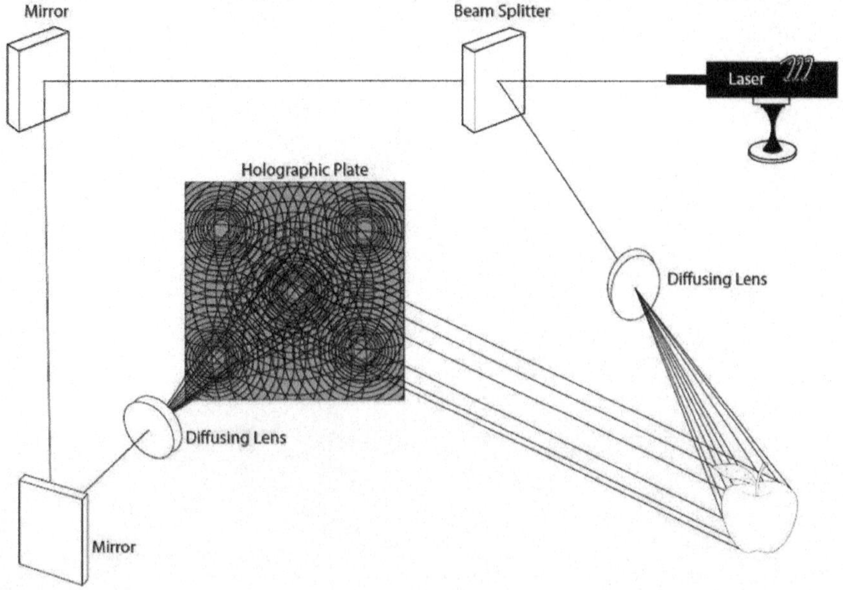

The secondary beam is reflected off of two mirrors before rejoining the first beam, creating the three dimensional appearing holographic image on the receiving plate. The reference beam is unaltered, direct from the source, and the secondary beam is a diffused reflection creating an energetic representation of the implicate/explicate order that parallels spiraling DNA strands.

Mirrors live up to their mythologized magical qualities in unexpected ways and the reflection that they manifest contains the nested implicate essence of a hologram. In true holographic fashion, the entire image can be seen in every piece of a broken mirror, demonstrating the implicate/explicate order like the proverbial drop of water in the ocean.

Fa-Tsang's analogy of the interconnectedness and interpenetration of all things holds true here. The whole is implicit in each of its parts and every point in the cosmos is its center. This vision of the universe as a multi-dimensional network of jewels, shows each one reflecting all others infinitely.

At the heart of it all the candle in the middle of a room full of mirrors shows the relationship of the One to the many, and the polished crystal in the center of the room reflecting everything around it shows the relationship of the many to the One.

This universal paradigm also plays itself out endlessly in the deepest levels of our psychological makeup. The people in our lives mirror our different aspects and sub-personalities, providing us with reflections of our inner world and the outer world as a reflection of the inner, much of which is shadow. Some of the reflections are beautiful and many are undesirable and denied.

The circular diagram from Robert McKee's *Story* provides an excellent map of the zones and levels of personal conflict we can experience. Any zone can interact with any other zone in any number of ways, providing ample opportunity to discover the complexity of our inner and outer relationships.

This can also be envisioned as the "center of our universe" as it maps the interplay of energetic interactions that start in the middle and work their way out.

At the center of our universe we find the innermost self which is composed of mind, body, and emotions. Everyone knows what it is like to have conflict between these three forces. Typically, one of them leads while the other two are relegated to shadow until a strong stimulus causes them to pop out, sometimes when least expected.

When confronted with something we can react first with our intellect or mind, because that is what we have grown to depend on, then we might act by doing something by moving our body, then we might feel good or bad about what we have done through emotion.

Someone else might react with their feelings first through emotion, then think about how they reacted with their mind, then act out using their body. Anyone can act in any combination of ways, but they typically lead with their strongest most habitual response at the expense of the other two.

What would it be like if someone reacted with all three of these forces simultaneously in a balanced way?

This is one of the goals of integrating the shadow in the process of what Jung called individuation, which is one of the rewards of personal power a shaman gets from getting psychologically dismembered or swallowed by the jaguar.

These three primary inner energies have been characterized throughout myth and history in many ways. The thinking mind is referred to as knowledge or wisdom, the moving body is referred to as energy or power, while the feeling of emotion is referred to as love. On one end of the spectrum the modern divide and conquer scientific mind set has given mind the most attention. Energetically it is the most stable. On the other end, emotion is the quickest, most mercurial, and hardest to control. The body usually serves the other two, though it holds deep wisdom of its own that is often ignored.

The Inca view of reality is a shamanic one of three worlds; the upper world, represented by the condor, which has a rose color and is thought of as love or emotion. The middle world is represented by the jaguar or puma, and has an electric blue color signifying power manifest in the body, and the lower world is represented by the serpent which is gold, signifying wisdom. When the three are combined in harmony they are believed to create a beautiful electric violet hue.

These three energies have been depicted in archetypal stories that dramatize secondary characters who mirror the inner life of the protagonist. One of the most striking examples of this is *The Wizard of Oz* where Dorothy is on her hero's quest to find her way home. Her three primary allies are the tin man who wants a heart as love or emotion, the cowardly lion, who wants courage as power acting through his body, and the scarecrow, who wants a brain to manifest wisdom through his mind.

Shamans navigate multiple realities within these infinite nested reflections knowing that their power lies in how they filter, see, and react to the realms that they come in contact with by manipulating their perception with the knowledge that all they encounter comes to them through their subjective experience.

In shamanism as in holograms, everything is connected to everything else. Plants, animals, man, and elemental spirits all interact and affect each other in visible and invisible ways. For them the center of their universe is right between their eyes, but their real home is in the heart.

Subjectively we are a veritable "cast of thousands" that can change moods, perceptions, and outlooks, depending on which sub-personality is in charge, and this can change from moment to moment. If we observe closely, we can see instantaneous changes in the facial expressions of people when their sub-personalities surface to take control, making them look like different people than the face they usually show to the world, particularly when they are stressed.

As far back as 1915 Gurdjieff characterized this phenomenon in a talk to his followers.

"One of man's important mistakes, one which must be remembered, is his allusion in regard to his I.

'Man such as we know him, the 'man machine', the man who cannot 'do', and with whom and through whom everything 'happens', cannot have a permanent and single I. His I changes as quickly as his thoughts, feelings, and moods, and he makes a profound mistake in considering himself always one in the same person; in reality he is *always a different person*, and not the one he was a moment ago.

'*Man has no permanent and changeable I.* Every thought, every mood, every desire, every sensation, says I. And then each case it seems to be taken for granted that this I belong to the *Whole*, to the whole man, and that a thought, a desire, or an aversion is expressed by this Whole. In actual fact there is no foundation whatever for this assumption. Man's every thought and desire appears and lives quite separately and independently of the Whole. And the Whole never expresses itself, for the simple reason that it exists, as such, only physically as a thing, and in the abstract as a concept. Man has no individual I. But there are, instead, hundreds and thousands of separate small I's, very often entirely unknown to one another, never coming into contact, or, on the contrary, hostile to each other, mutually exclusive and incompatible. Each minute, each moment, man is saying or thinking 'I'. And each time his I is different. Just now it was a thought, now it is a desire, now a sensation, now another thought, and so on, endlessly. *Man is a plurality.* Man's name is legion.

'The alteration of I's, their continual obvious struggle for supremacy, is controlled by accidental external influences. Warmth, sunshine, fine weather, immediately call up a whole group of I's. Cold, fog, rain, call up another group of I's, other associations, other feelings, other actions. There is nothing in man able to control this change of I's, chiefly because man does not notice, or know of it; he lives always in the last I."

We learn our coping strategies by mirroring the behaviors of those around us, which is why Gurdjieff said that we are "controlled by accidental external influences", especially when dealing with traumatic situations. Although these deeply embedded behaviors were learned by our subpersonalities in the same manner, this does not make them correct responses. More often than not, these learned responses are replete with errors of interpretation and reaction that can temporarily defuse the trauma happening in the moment at hand, but do not solve the actual problem in the long run. Ultimately these responses are delegated to the depths of shadow in the form of sub-personalities that become trauma induced stock responses acting as protective mechanisms of illusion between the ego and the real world.

The important point to grasp here is that *we are the creators of these traumatized entities that fill our inner personal universe reflected by those outside of us*, especially when we project them outside of us and deny their existence within us, essentially abandoning our own creations.

When we meet people for the first time, we seek out what are referred to as "attention points", which are things we have in common with them like sports, music, the arts, or anything else we can connect with that is a mutual interest. In this manner we pave the way to sharing similar thoughts and seek resonance with that person by sharing common vibrations as a place and energy to bond. This whole phenomenon is especially prevalent in present day tribal rave and festival culture where large groups of people bond in a loving vibration, often enhanced by drugs like MDMA which is considered an empathogen along with a class of psychoactive drugs referred to as entheogens that produce experiences of emotional communion, oneness, relatedness, empathy, and sympathy. These experiences bond them in the shared medium of the music and the tribal beats, enveloping them in an all-around "feel good vibe."

What we don't realize in our bonding and meeting of people is that the subconscious elements of our shadow also draws them to us through its familiarity and recognition of their shadow aspects.

When we find ourselves swallowed up in negative emotions they tend to overwhelm and blind us to the truth of the situation through judgment, projection, denial, and other egoic survival strategies, but if we can "wake up" to the illusion and raise our awareness to the point of having a greater presence of mind we can ask important questions of our reactions.

What is the quality of this emotional energy and how does it make us feel?

Whose eyes are we seeing it through?

What mask are we wearing?

What filters have we taken on?

All of this interconnectedness and interaction reinforces the notion that we are all one, connected in our hearts, minds, and spirits. As mentioned, shamans believe everything is interconnected, including man, plants, animals, trees, and every other part of creation whether manifesting as visible or invisible energies. American Indians acknowledge this sacred universal order and refer to these elemental beings as brothers.

AN INFINITE HALL OF MIRRORS

If you look at the internet with this interconnected mirroring paradigm in mind, you will realize that the World Wide Web is a huge holographic mirror of the collective consciousness of humanity with all of the information contained everywhere in the whole. Like our personal shadow aspects, much of this information is not accurate. It comes from someone's reflection on something which is often distorted and biased either unconsciously or intentionally.

An interesting property of mirrors is that they can show us what we do not want to see and if the mirror is distorted, so is the reflection.

In World War II before television, those outside of the war zone got their information from newspapers, radios, and newsreels, all filtered through the individual viewpoints of reporters, editors, and government censors. The Vietnam War was called the first televised war, but even those images were filtered through a few major networks censored by their editors and the government.

In the twenty-first century we have internet access with technological mirrors known as digital cameras, cell phones, and all manner of instantaneous communication. Editors and censors are being bypassed because of these mirrors and the mirror of truth is laying bare the lies of governments, the atrocities of war, and the lies and propaganda of diverse agendas, but the realities of these infinite agendas make it harder to discern where the real truth lies.

Everyone is buzzing about the latest scandals, incidents of police brutality, racial conflicts, mass murders, political corruption and more. The list is endless. The burgeoning phenomena of social media allowed Egyptians to execute a relatively nonviolent overthrow of an oppressive government, and what violence did occur was instantly transmitted out to the World Wide Web for the entire world to see.

Literally on another front, many believe that our erratic weather patterns, floods, wildfires, and earthquakes are a reflection of the turbulence going on within human consciousness, and many believe that this turbulence is the chaos that precedes the birth of something new.

We are all connected and each of us is a mirror to each other from the macrocosm of the collective, down to the microcosm of our individual selves, to the sub-personalities that make up what we think of as "I", but is in fact many. Some of our sub-personalities we like and constantly take them out to show to the world. Others we don't like we repress, making them shadow that we don't want to acknowledge.

Although mirrors often show us what we do not want to see. If you have the ability to be honest with yourself and desire to be an integrated person, then the work lies in reintegrating your hidden shadow(s) which are cunning and elusive. Some would rather see the death of you than be found out, and they think that if they are discovered they will die, which contains a grain of truth because in order to be reborn, a death is necessary.

One of the most used strategies that the shadow employs is to project itself onto others in self-righteousness and judgment to muddy the waters and distort the reflection. If you can bear the reflection you will realize that what you strongly dislike in others creates strong emotional reactions because it is a reflection of what you do not want to acknowledge or take responsibility for inside of you. It takes great courage to look at yourself in the glare of this mirror.

The mirrors are reflecting back from everywhere that you look, and any time you feel self-righteous, judgmental, and superior, you can be sure that your shadow is being reflected back to you.

Once you discover and embrace these abandoned, infantile ego creations with compassion, you can acknowledge what you created. In spite of their misguided actions they are trying to protect you at all costs. You reside at the center of their universe, so take a good look at who is in the "outer universe" of your life and the feelings that come from the reflections they engender in you.

The infinite reflections nested within this holographic paradigm was mapped out in ancient times in the form of the Temple of Anthropocosmic Man which was architecturally rendered to exhibit the proportions of man within its design and artwork as well as the mathematical and geometrical structure of the Cosmos and its locale within human consciousness. Pharaonic Consciousness recognized man as the center of the Universe and formally equated it within this structure.

Gurdjieff had a similar interpretation.

"There is no need to study or investigate the sun in order to discover the matter of the solar world: this matter exists in ourselves and is the result of the division of our atoms. In the same way we have in us the matter of all other worlds. Man is, in the full sense of the term, a 'miniature universe'; in him are all the matters of which the universe consists; the same forces, the same laws that govern the life of the universe, operate in him; therefore in studying man we can study the whole world, just as in studying the world we can study man."

Gurdjieff's teachings followed the same principles as the Temple of Man, although his diagram is different and includes All Worlds and the Moon in what he called "The Ray of Creation", characterized as an octave that repeats at many levels and sub-levels.

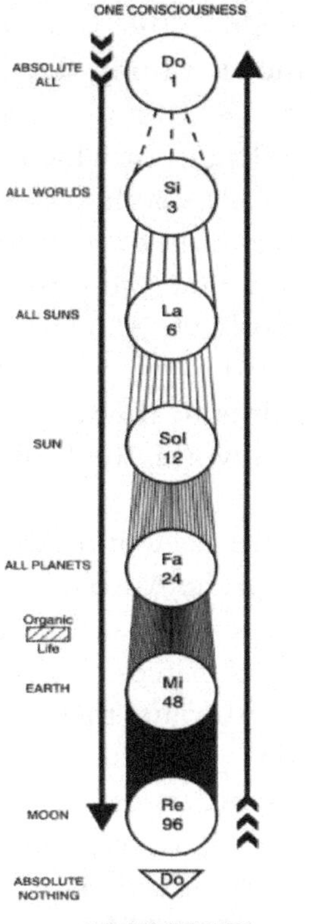

A 'RAY OF CREATION'
showing the multiplying orders of laws and the increasing density of materiality at each level.

In Gurdjieff's model, the Whole, One, or *All*, he called the "Absolute" because including everything within itself, it is not dependent upon anything. It is the "world" for "all worlds" where All forms one single Whole, infinite and indivisible as the primordial state of things out of which by division and differentiation arises the diversity of the phenomena observed by us. In the Absolute, vibrations are the fastest, and matter is the least dense, and in the descending ray of creation, the dampening vibrations are slower, and matter is more dense.

Gurdjieff stated that matter possessing characteristics of materiality comprehensible to us is divided into several states according to its density; solid, liquid, and gaseous. Further gradations of matter are radiant energy like electricity, light, magnetism, and so on.

He referred to this octave as the table of hydrogen that determined the density of matter and the speed of vibrations, as well as the degree of intelligence and consciousness. The degree of consciousness corresponds to the degree of density or the speed of vibrations meaning the denser the matter, the less conscious and the less intelligent, and the denser the vibrations, the more conscious and the more intelligent.

The chain of worlds, the length of which are the Absolute forms the "ray of creation", which for us is the "world" in the widest sense of the term. The Absolute gives birth to an infinite number of different worlds, each of which begins a new and separate ray of creation. The number of forces in each world, 1, 3, 6, 12, and so on indicate the number of laws the given world is subject to. The fewer laws that are in a given world, the nearer it is to the will of the Absolute, and the more laws there are in a given world, the further it is from the will of the Absolute.

Gurdjieff taught that we live in a world subject to forty-eight orders of laws very far from the will of the Absolute and in a very remote and dark corner of the universe. The ray of creation from world 1 down to world 96 indicate the number of forces, or orders of laws that govern the worlds in question. In the Absolute there is only one force and one law - the single and independent will of the Absolute. In the next world there are three forces or three orders of laws, in the next six orders of laws, and in the following one, twelve, and so on. According to Gurdjieff, the earth has forty-eight orders of laws that we are subject to. If we lived on the moon we would be subject to ninety-six orders of laws and would not have the possibilities of escape we now have.

Following this octave model of increasing densities of matter and creation as well as intelligence, Gurdjieff defined the end of the ray, saying, "The souls that go to the moon, possessing perhaps even a certain amount of consciousness and memory, find themselves there under ninety-six laws, in the conditions of mineral life, or to put it differently, in conditions from which there is no escape apart from a general evolution in immeasurably long planetary cycles. The Moon is "at the extremity", at the end of the world; it is the "outer darkness" of the Christian doctrine "where there will be weeping and gnashing of teeth".

This ancient depiction of a holographic world view with the microcosm in the macrocosm demonstrated in the Temple of Anthropocosmic Man and in Gurdjieff's diagram represent a shamanic belief that infuses the shaman's transformation and the ancient archetype of The Hero's Journey. In the case of the temple, its design, layout, and its mathematical meaning provide the framework for our inner and outer universe.

In this model the heart is considered the giver of life located at the center of man. It gives life to and supports the other organs surrounding it in the same way that the sun at the center of our solar system gives life to and supports the planets surrounding it. Each planet that orbits this solar furnace which is the crucible of the heart has historically been attributed with its own personality. Mars has been called "The God of War", Venus "The Goddess of Beauty and Love". The others each have their own unique attributes, meaning their own unique energies.

The Solar System

Centuries before it became a pop culture fad, astrology was an integral part of astronomy. In addition to the earth and the seasonal signs of the zodiac, astrology deals with ten planets, namely the Sun and the Moon, the two luminaries that are considered planets in astrology, Mercury, Venus, Mars, Jupiter, Saturn, Uranus, Neptune, and Pluto. Each planet has its precise function in the natal chart and represents a specific energy. The planet's action is influenced by the sign its tenets and primarily plays out in the area of life indicated by the house it is posited in.

Like the Temple of Anthropocosmic Man, in shamanism the heart in man is the sun of his personal cosmos that is connected to the sun which is the giver of life in our solar system, especially on planet earth. This sun is connected to a bigger sun, which is connected to a bigger sun, infinitely following Gurdjieff's ray of creation all the way back to Source.

Hafez, a 14th-century poet, best known for expressing the ecstasy of divine inspiration in the **mystical** form of love poems penned a verse that captures the essence of this belief.

"Even After All this time The Sun never says to the Earth, 'You owe me.' Look What happens With a love like that, It lights the whole sky."

An old adage in shamanism made popular by Carlos Castaneda attributed to Don Juan, the Yaqui shaman, is that, "A warrior must follow a path with heart." In other words, follow your heart. It will take you back to Source and you will have found your way home.

In the same way that the sun at the center of our solar system is influenced by and influences the planets in its orbit, so do the organs and their energies in the human body interact with its heart.

We are all at the heart of the universe that we are creating. Like the planets, the people in our lives are energetic influences that represent different aspects of us that manifest different sides of our personalities. Our supporting allies mirror and support the good, including positive aspects of our shadow, and the antagonistic ones reflect our darker side.

In visionary states information is communicated internally through alien feeling symbols, concepts, emotions, thoughts, vistas, and other mixed, often synesthetic perceptions containing dense information that often unfolds through transforming geometric colors and patterns.

Not only do these mesmerizing multi-hued geometric mathematical permutations constitute a unique language primarily available to the inner perception of the human mind, but geometry in and of itself is a universal language containing self-apparent truths, regardless of any differences in the spoken language of the perceivers.

Galileo asserted that only those properties of matter that are directly amenable to mathematical measurement like size, shape, and weight are real, while the other more subjective qualities like sound, taste, and color are illusionary impressions, since the book of nature is written in the language of mathematics alone. In his words:

"The grand book the universe... is written in the language of mathematics and its characters are triangles, circles, and other geometric figures without which it is humanly impossible to understand a single word of it; without these, one wanders about in a dark Labyrinth."

Though Galileo may have missed the mark regarding the measurement of the "subjective qualities" of sound, taste, and color in light of modern day methods that measure what he could not perceive with the technology of his day, there is great truth in his comments about the language of mathematics and geometry which shamans have always known as the heart centered universal language that they are.

THE HEART OF THE MATTER

The human heart pumps life-giving oxygenated nutrient-rich blood throughout our bodies on a precise schedule. Researchers are learning that this extraordinary machine weighing on average less than 10 ounces, also possess a level of intelligence they are only beginning to understand. Evidence shows the heart also plays a greater role in our mental, emotional, and physical processes than previously thought.

HeartMath Institute Director of Research Rollin McCraty wrote in his paper, *The Energetic Heart: Bioelectromagnetic Communication Within and Between People* that, "The heart is a sensory organ and acts as a sophisticated information encoding and processing center that enables it to learn, remember, and make independent functional decisions." Scientists at the HeartMath Institute like McCraty have been conducting research for over two decades that sheds light on what researchers now characterize as the *energetic heart.*

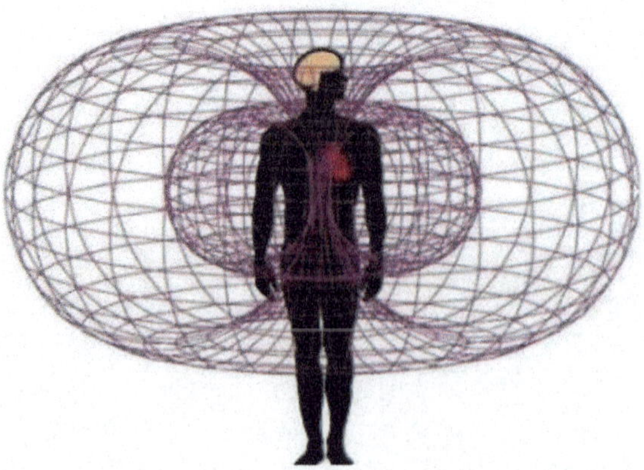

Like the brain, the heart generates a powerful electromagnetic field, McCraty explains. "The heart generates the largest electromagnetic field in the body. The electrical field as measured in an electrocardiogram (ECG) is about 60 times greater in amplitude than the brain waves recorded in an electroencephalogram (EEG)."

HeartMath studies show that this powerful electromagnetic field can be detected and measured several feet away from a person's body and between two individuals in close proximity.

In the HeartMath study, *The Electricity of Touch: Detection and Measurement of Cardiac Energy Exchange Between People*, researchers set out to determine whether the heart's electromagnetic field as measured by an electrocardiogram in one individual can be detected and measured in another person when the pair were seated within about three feet of each other or held hands.

The results of *The Electricity of Touch* experiment showed that when people touch or are in proximity, a transference of the electromagnetic energy produced by the heart occurs. The authors observed important implications raised by this research when viewed in conjunction with the success of numerous healing modalities.

Practices like therapeutic touch, holoenergetic healing, healing touch, chi gong and reiki among others are based on the assumption that an exchange of energy occurs to facilitate healing according to the findings. While scientific evidence exists to substantiate the physiological and psychological effects of many of these treatments, science has not been able to describe a mechanism that this putative energy exchange between individuals takes place within. This study, together with the work of other researchers represents *one* of the first successful attempts to directly measure an exchange of energy between people.

The heart emits an electrical field *60 times greater in amplitude* than the activity in the brain, and an electromagnetic field 5,000 times stronger than that of the brain. From an electromagnetic perspective, *the heart's electromagnetic field is 5,000 times more powerful than the brain's.*

Heart intelligence is the flow of intuitive awareness, understanding, and inner guidance we experience when the mind and emotions are brought into coherent alignment with the heart. The human heart, the size of two adult fists, is mysterious, intelligent, powerful, and often inexplicable.

The Egyptians believed that Anubis, the god of the underworld and judge of the dead, weighed the hearts of the recently deceased against a feather. If the two balanced the heart would be returned to the owner. If the heart was heavier it was weighted by bad deeds and fed to a monster.

Our hearts beat at twice the pace of most animals and humans have vascular disease, but our cousins the great apes do not. With its own electrical impulse, the human heart can continue to beat if taken from the body. Beating heart cells grown in Petri dishes synchronize in resonance with each other and the heart emits a signature electrical frequency thousands of times more powerful than anything else in the body.

Neurons, the brain cells responsible for processing sense-based input, send messages to the body like, "reach the hand to pick up a sandwich." Neurons also transmit emotion. These specialized cells are found in the brain and nervous system, but also in the heart and can be harnessed to establish heart-brain coherence. In fact, heart neurons fire in conjunction with brain neurons and the heart and brain are connected in profound ways.

Heart intelligence is the flow of awareness, understanding, and intuition we experience when the mind and emotions are brought into alignment with the heart. It can be activated through self-initiated practice, and the more we pay attention when we sense the heart is speaking to or guiding us, the greater our ability to access this intelligence and guidance. Heart intelligence underlies cellular organization and guides and evolves organisms toward increased order, awareness, and coherence of their bodies' systems.

Throughout history human beings have understood that *intelligence*, the ability to learn, understand, reason, and apply knowledge to shape their environment was a function of the brain. There is ample evidence in the writings and oral traditions passed down through generations that also believed in an intelligent heart.

Research into the idea of heart intelligence accelerated in the second half of the 20th century. During the 1960s and 70s pioneer physiologists John and Beatrice Lacey conducted research that showed the heart communicating with the brain in ways that greatly affect how we perceive and react to the world around us. In 1991, the year the HeartMath Institute was established, pioneer neurocardiologist Dr. J. Andrew Armour introduced the term "heart brain," stating that the heart possessed a complex and intrinsic nervous system that is a brain. Today, more than a half century after the Laceys began their research, we know a great deal more about the heart:

- The heart sends us emotional and intuitive signals to help govern our lives.
- The heart directs and aligns many systems in the body so they can function in harmony with one another.
- The heart is in constant communication with the brain. The heart's intrinsic brain and nervous system relay information back to the brain in the cranium, creating a two-way communication system between heart and brain.
- The heart makes many of its own decisions.
- The heart starts beating in an unborn fetus before the brain has formed, a process scientists call autorhythmic.
- Humans form an emotional brain long before a rational one, and a beating heart before either.
- The heart has its own independent complex nervous system known as "the brain in the heart."

"Researchers began showing in the 1980s and 90s that success in life depended more on an individual's ability to effectively manage emotions than on the intellectual ability of the brain in the head," says *The HeartMath Solution*, by HeartMath founder Doc Childre and his longtime HeartMath spokesman Howard Martin.

This discovery resulted in people wanting to know how to infuse emotions with intelligence. Scientists at the nonprofit HeartMath Institute conducting research into heart intelligence and emotions posed the theory that heart intelligence transfers intelligence to the emotions and instills the power of emotional management, making it the source of emotional intelligence.

These scientists state, "From our research at the HeartMath Institute, we've concluded that intelligence and intuition are heightened when we learn to listen more deeply to our own heart. It's through learning how to decipher messages we receive from our heart that we gain the keen perception needed to effectively manage our emotions in the midst of life's challenges. The more we learn to listen to and follow our heart intelligence, the more educated, balanced and coherent our emotions become. Without the guiding influence of the heart we easily fall prey to reactive emotions such as insecurity, anger, fear and blame as well as other energy-draining reactions and behaviors."

Early HeartMath research found that negative emotions threw the nervous system out of balance and when that happened heart rhythms became disordered and jagged on a heart monitor placing stress on the physical heart and other organs threatening serious health problems.

"Positive emotions, by contrast, were found to increase order and balance in the nervous system and produce smooth, harmonious heart rhythms," Childre and Martin wrote, "But these harmonious and coherent rhythms did more than reduce stress: They actually enhanced people's ability to clearly perceive the world around them."

Aristotle considered the heart as the center of reason, thought, and emotion, senior to the brain in importance. Ninth century Arabic philosopher Abu Nasr al-Farabi believed that, "The ruling organ in the human body is the heart; the brain is a secondary ruling organ subordinated to the heart." Auguste Comte, a 19th century French philosopher declared that the brain should be servant to the heart.

"The most common denominator in all religions is that the heart is the seat of wisdom," said Rollin McCraty Ph.d, director of research at the groundbreaking HeartMath Institute in Santa Cruz, CA. Twelfth century Christian mystic, Hildegard of Bingen wrote, "The soul sits at the center of the heart, as though in a house."

The heart's electromagnetic field contains information that researchers are trying to understand which is transmitted throughout and outside of the body. One of the most significant findings of HMI's research related to this field is how intentionally generated positive emotions can change this information and coding.

That discovery raises questions about whether the cardioelectromagnetic field information transmitted from an individual who is angry, fearful, depressed, or experiencing some other negative emotion takes on beneficial properties when it is influenced by positive emotions. Also, is the care, compassion, love or other positive emotion not only transmitted throughout an individual's body as the cardioelectromagnetic field radiates through it, but externally as well to people in close proximity or over long distances?

Although HeartMath continues studying the effects of human electromagnetic fields over large distances, researchers say the answer to the first question, whether self-regulated positive emotions can be beneficial to the individual, is yes.

According to the paper *Emotional Stress, Positive Emotions, and Psychophysiological Coherence*, written by HMI researchers McCraty and Dana Tomasino, "Through the use of tools and technologies that foster positive emotions and psychophysiological coherence, individuals can effectively initiate a repatterning process, whereby habitual emotional patterns underlying stress are replaced with new, healthier patterns that establish increased emotional stability, mental acuity, and physiological efficiency as a new familiar baseline or norm."

Additional research is needed to determine the nature and function of this energy/information the heart sends out. HMI's researchers have conducted numerous studies on how positive emotions can affect us. Participants in several studies, and thousands of people worldwide using the HeartMath System of tools and technology can attest to a broad range of benefits resulting from intentionally experiencing certain emotions, which HeartMath calls emotion self-regulation.

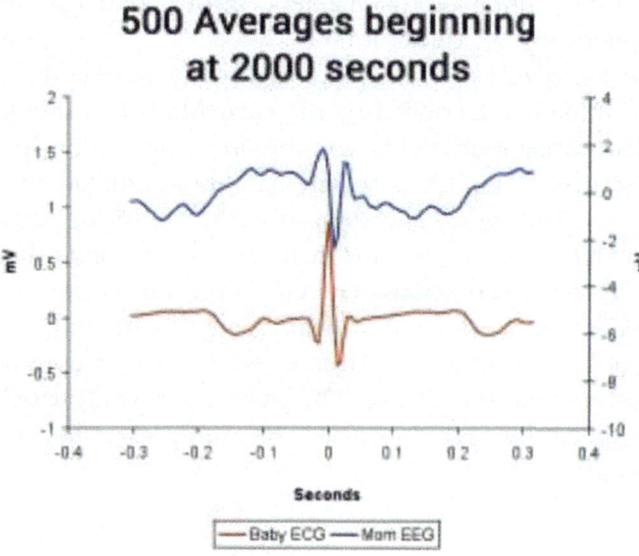

The top wave in the above graph is a mother's signal averaged electroencephalograph, or brainwave signal. The bottom wave is her infant's signal averaged electrocardiogram, or heart signal. Signal averaging is a technique for detecting synchronization between two signals. In this example, the mother's brainwave is synchronized to her baby's heartbeat.

In another study, researchers examined whether there is an energy exchange, in this case heart-brain interactions, in one of the most intimate of human relationships. The Proof of Concept Study: *Heart-Brain Synchronization between Mother and Baby*, explored the potential to measure energetic heart-brain interactions occurring between a mother and her infant.

Researchers showed that the mother's brainwaves synchronized to her baby's heartbeat with the baby lying in the mother's lap with a blanket placed in between mother and baby. In the summary of their findings, the study's authors wrote, "This preliminary data elucidates the intriguing finding that the electromagnetic signals generated by the heart have the capacity to affect others around us. It appears that when the mother placed her attention on the baby that she became more sensitive to the subtle electromagnetic signals generated by the infant's heart. These findings have intriguing implications, suggesting that a mother in a psychophysiologically coherent state became more sensitive to the subtle electromagnetic information encoded in the electromagnetic signals of her infant."

One of the questions being explored by the science-based Global Coherence Initiative, launched by the HeartMath Institute is whether human electromagnetic fields can be influenced non-locally over greater distances. The GCI, which has about 25,000 international participants is utilizing advanced magnetometer sensing technology to observe changes in the earth's magnetic field by, among other possible effects, mass human emotions. The GCI's greater purpose is "to unite millions of people worldwide in heart-focused care and intention, to help shift global consciousness from instability and discord to balance, cooperation and enduring peace. The HeartMath Institute believes that this purpose will be served as scientists gain greater insight into the energetic heart.

YOUR HEART'S INFLUENCE IS GREATER THAN YOUR HEAD'S

We have a "heart brain" with a vast array of neurons that process sensory information on their own and communicate that information to our brain, nervous system, and other organs, meaning the heart can learn, remember, and make decisions independent of the brain. Proof of this comes from heart transplant cases where the person who received the heart has memories, experiences, and emotions of their deceased donors.

According to the HearthMath Institute, the heart sends more signals to the brain than the brain sends to the heart. Numerous experiments have demonstrated that the signals the heart sends to the brain influence the function of higher brain centers involved in perception, cognition, and emotional processing. Additionally, and the heart generates the body's most powerful electromagnetic field that is 100x stronger electrically, and 1000x stronger magnetically than the brain's. Through its neural electrical and magnetic fields, the heart has the most powerful ability to take in large amounts of sensory information that influences our brain and internal system. It also has the ability to influence our surroundings with its expansive energetic/magnetic field, making the heart a major resource for information, wisdom, and influence over our internal and external experiences.

The communication between the heart and brain is a dynamic, ongoing, two-way dialogue with each organ continuously influencing the other's function. Research has shown that the heart communicates to the brain in four major ways:

- *neurologically* through the transmission of nerve impulses
- *biochemically* via hormones and neurotransmitters
- *biophysically* through pressure waves
- *energetically* through electromagnetic field interactions

Communication along all these conduits affects the brain's activity, and messages that the heart sends to the brain affect performance.

During 20 years of research throughout the 1960s and 70s, John and Beatrice Lacey observed that the heart communicates with the brain in ways that significantly affect how we perceive and react to the world.

In researcher Walter Cannon's view, when we are aroused the mobilizing part of the nervous system referred to as the sympathetic energizes us for fight or flight, indicated by an increase in heart rate. In more quiescent moments the calming part of the nervous system referred to as the parasympathetic calms us down and slows the heart rate. Cannon believed the autonomic nervous system and all of the related physiological responses moved in concert with the brain's response to any given stimulus or challenge. In his view all of our inner systems are activated together when we are aroused and calm down together when we are at rest with the brain in control of the entire process.

Cannon introduced the concept of homeostasis which is the study of physiology based on the principle that all cells, tissues, and organs strive to maintain a static or constant steady-state condition. With the introduction of signal-processing technologies that can acquire continuous data over time from physiological processes like heart rate, blood pressure, and nerve activity, it has become apparent that biological processes vary in complex and nonlinear ways, even during so-called steady-state conditions.

These observations led to the understanding that healthy, optimal function is a result of continuous, dynamic, bidirectional interactions among multiple neural, hormonal, and mechanical control systems at both local and central levels. These dynamic and interconnected physiological and psychological regulatory systems never rest and are never static. This is also evident in the normal resting *rhythm* of the heart which is variable rather than regular.

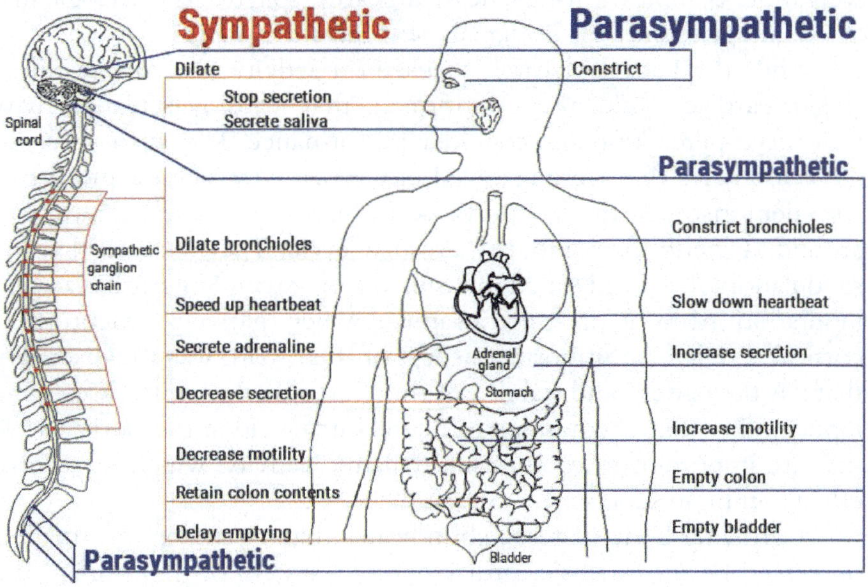

Parasympathetic fibers are primarily in the vagus nerves, but some that regulate subdiaphragmatic organs travel through the spinal cord. Sympathetic fibers also travel through the spinal cord. Emotions can affect activity in both branches of the autonomic nervous system. Anger causes increased sympathetic activity while relaxation increases parasympathetic activity.

The Laceys noticed that Cannon's model only partially matched physiological behavior and discovered that the heart appeared to have its own logic that often diverged from the direction of autonomic nervous system activity with the heart behaving as if it had a mind of its own.

The heart appeared to be sending meaningful messages to the brain that the brain understood and obeyed with messages that affected perception, behavior, and performance. The Laceys identified a neural pathway and mechanism with input from the heart to the brain that could inhibit or facilitate the brain's electrical activity.

In 1974, French researchers stimulated the vagus nerve that carries many of the signals from the heart to the brain in cats and found that the brain's electrical response decreased to about half its normal rate, suggesting that the heart and nervous system were not simply following the brain's directions. This indicated that the autonomic nervous system and the communication between the heart and brain were more complex, with the heart appearing to have its own logic that acted independently of the signals sent from the brain.

While the Laceys research focused on activity occurring within a single cardiac cycle they confirmed that cardiovascular activity influenced perception and cognitive performance. They also observed inconsistencies that were resolved in Germany by Velden and Wölk, who demonstrated that cognitive performance fluctuated at a rhythm around 10 hertz throughout the cardiac cycle. They showed that the modulation of cortical function resulted from ascending cardiovascular inputs on neurons in the thalamus, which globally synchronizes cortical activity. An important aspect of their work was the discovery that it is the pattern and stability of the heart's rhythm of the ascending inputs rather than the number of neural bursts within the cardiac cycle that are important in modulating thalamic activity, which has global effects on brain function.

A growing body of research indicates that ascending information processed by the intrinsic cardiac nervous system of the heart to the brain can influence activity in the frontocortical areas and motor cortex affecting psychological factors like attention level, motivation, perceptual sensitivity, and emotional processing.

These findings inspired a small group of cardiologists, neurophysiologists, and neuroanatomists to create a new discipline called *neurocardiology*. One of their early findings is that the heart has a complex neural network that is sufficiently extensive to be characterized as a brain in the heart. The *heart-brain*, or intrinsic cardiac nervous system, is an intricate network of complex ganglia, neurotransmitters, proteins, and support cells similar to those of the brain in the head. The heart-brain's neural circuitry enables it to act independently of the cranial brain to learn, remember, make decisions, feel, and sense. Descending activity from the brain in the head via the sympathetic and parasympathetic branches of the autonomic nervous system are integrated into the heart's intrinsic nervous system along with signals arising from sensory neurons in the heart that detect pressure, heart rate, heart rhythm, and hormones.

The anatomy and functions of the intrinsic cardiac nervous system and its connections with the brain have been explored by neurocardiologists. In terms of heart-brain communication, it is well-known that the descending pathways in the autonomic nervous system are involved in the regulation of the heart, but it is less appreciated that the majority of fibers in the vagus nerves are ascending in nature. More of these ascending neural pathways are related to the heart and cardiovascular system than to any other organ, meaning the heart sends more information to the brain than the brain sends to the heart, providing further proof that the neural interactions between the heart and brain are more complex than previously thought. In addition, the intrinsic cardiac nervous system has both short-term and long-term memory functions that operate independently of central neuronal command.

Once information has been processed by the heart's intrinsic nervous system signals are sent to the heart's sinoatrial node and to other tissues in the heart. Under normal physiological conditions the heart's intrinsic nervous system plays an important role in the routine control of cardiac function, independent of the central nervous system. The heart's intrinsic nervous system is also vital for the maintenance of cardiovascular stability and efficiency, and the heart cannot function properly without it. The neural output or messages from the intrinsic cardiac nervous system travel to the brain through ascending pathways in the spinal column and vagus nerves where it travels to the medulla, hypothalamus, thalamus, and amygdala, then to the cerebral cortex.

In addition to its extensive neurological interactions, the heart also communicates with the brain and body biochemically through hormones it produces. Although not thought of as an endocrine gland, the heart manufactures and secretes a number of hormones and neurotransmitters that have a wide-ranging impact on the body as a whole.

The heart was reclassified as part of the hormonal system in 1983 when a new hormone produced and secreted by the atria of the heart was discovered. Called by several different names like the atrial natriuretic factor, atrial natriuretic peptide, and atrial peptide, it was nicknamed the balance hormone. It plays an important role in fluid and electrolyte balance and helps regulate the blood vessels, kidneys, adrenal glands, and many regulatory centers in the brain. Increased atrial peptide inhibits the release of stress hormones, reduces sympathetic outflow, and appears to interact with the immune system. Experiments suggest that atrial peptide can also influence motivation and behavior.

The heart also contains cells that synthesize and release catecholamines like norepinephrine, epinephrine, and dopamine, neurotransmitters originally thought to be produced only by neurons in the brain and ganglia. The heart also manufactures and secretes oxytocin which can act as a neurotransmitter, and is commonly referred to as the love or social bonding hormone. Beyond its well-known functions in childbirth and lactation, oxytocin has been shown to be involved in cognition, tolerance, trust, friendship, and the establishment of enduring pair-bonds. Concentrations of oxytocin produced in the heart are in the same range as those produced in the brain.

Stress can adversely affect the physical functioning of the heart which can affect the cognitive and emotional functions of the brain. During stress and negative emotions heart rhythm patterns become erratic and disordered which sends neural signals to the brain to inhibit higher cognitive functions resulting in not being able to think clearly, remember, learn, reason, and make effective decisions. These signals then influence the emotional processes of the brain to go into a stress pattern as well.

Through the heart's electric and magnetic field, information is not only communicated throughout the body and brain, but to the external environment as well which is why a strongly negative person has the ability to influence our state and a loving person can shift us into a positive state. Our hearts have a huge magnetic field of energy capable of influencing those around us.

In positive emotional states the heart produces a more ordered stable pattern that sends inputs to the brain of positive feelings and emotional stability, creating better cognitive functioning.

Feeling good results in clearer thinking, feeling more motivated, and learning better. Practicing positive emotions and feeling them from the heart also creates a more stable heart pattern that significantly affects how you feel, think, and perform, mentally and physically.

The heart starts beating in an unborn fetus before the brain forms and is more like a second brain with the ability to make decisions irrespective of your "thinking" brain, and it can override your thinking and emotional brain. The heart also directs and aligns many systems in the body so they can function in harmony with each other. Finally, intuition arises from the heart and is considered the place where authenticity and the truth of who you are comes from. From this perspective it makes sense to allow the intelligence of your heart to be the driving force of your life as it is from this place that order, knowing, peace, love, and creation arise from.

Leading from the heart in a way that has positive effects internally on your mind, emotions, and body, and externally on those around you requires entraining your heart to ordered rhythmic patterns that create more powerful electric and magnetic fields of influence. This can be done by engaging in heart-centered activities like self-compassion, self-forgiveness, compassion and forgiveness for others, opening your heart, and practicing more heart connections with people as well as cultivating gratitude.

The heart plays a central role in creating physiological coherence and is associated with positive emotions and intuition. One of the strongest threads uniting diverse cultures and religious and spiritual traditions throughout history has been a universal regard of the heart as the source of love, wisdom, intuition, courage, and more. Everyone is familiar with expressions like "put your heart into it," "learn it by heart", "speak from your heart", and "follow a path with heart". These sentiments indicate that the heart is more than a physical pump that sustains life and reflect what is called the intuitive or spiritual heart. Throughout history people have turned to the intuitive heart, referred to as their inner voice, soul, or higher power, as a source of wisdom and guidance.

The terms intuitive and spiritual heart refer to our *energetic heart* which is coupled with a deeper part of ourselves referred to as our higher self, higher capacities, or what David Bohm described as "our implicate order and undivided wholeness." The term *energetic systems* in this context refers to functions we cannot directly measure, touch, or see, like emotions, thoughts and intuitions. Although these functions have loose correlations with biological activity, they remain covert and hidden from direct observation. Scientists have proposed that such functions operate primarily in the frequency domain outside of time and space, and have suggested possible mechanisms that govern how they interact with biological processes.

Our experience suggests that the physical heart has communication channels connecting it with the energetic heart. Nonlocal intuition is transformational and contains the wisdom that streams from the soul's higher information field down into the psychophysiological system through the energetic heart and can inform our moment to moment experience and interactions in what is referred to as *heart intelligence.* This is defined as the flow of higher awareness and the intuition experienced when the mind and emotions are brought into synchronistic alignment with the energetic heart.

When we are heart-centered and coherent we have a tighter coupling and closer alignment with our deeper source of intuitive intelligence and are able to more intelligently self-regulate our thoughts and emotions. Over time this lifts consciousness and establishes a new internal physiological and psychological baseline through an increased flow of intuitive information communicated by the emotional energetic system to the mind and brain systems resulting in a stronger connection with our deeper inner voice.

Our feelings result from our choices and we can change the way we feel by changing our thoughts. Positive feelings have a larger impact than we might imagine. These are all characteristics of consciousness, and as quantum physics shows, consciousness plays a definite role in the creation of our reality, and how we feel about things plays a role in this.

According to HeartMath Director of Research Dr. Rolin McCratey, "Emotional information is actually coded and modulated into these fields. By learning to shift our emotions, we are changing the information coded into the magnetic fields that are radiated by the heart, and that can impact those around us. We are fundamentally and deeply connected with each other and the planet itself."

A fundamental conclusion of quantum physics acknowledges that the observer creates the reality forcing physicists to admit that the universe is a "mental" construction. Pioneering physicist Sir James Jeans wrote: "The stream of knowledge is heading toward a non-mechanical reality; the universe begins to look more like a great thought than like a great machine. Mind no longer appears to be an accidental intruder into the realm of matter, we ought rather to hail it as the creator and governor of the realm of matter. Get over it, and accept the inarguable conclusion. The universe is immaterial-mental and spiritual."

INFINITE REFLECTIONS FROM THE HOLOGRAPHIC HEART BACK TO SOURCE

Changes in solar and geomagnetic activity correlate with human biology and are measured by autonomic nervous system activity. These energy fields impact us, the people around us, and every living thing connecting us in ways similar to quantum entanglement.

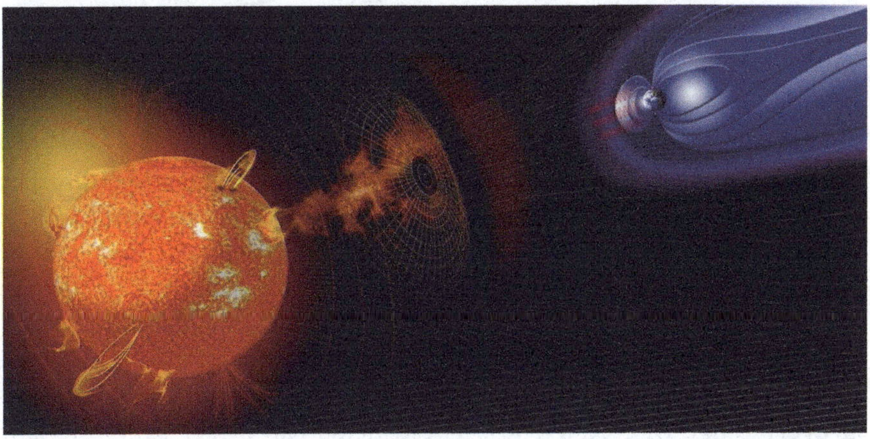

Energetically nested on a larger scale within the holographic order, the human body responds to changes in solar and geomagnetic activity.

Research suggests that daily autonomic nervous system activity responds to changes in solar and geomagnetic activity and synchronizes with the time-varying magnetic fields associated with geomatic field-line Schumann resonances.

In 1952, German physicist and professor W.O. Schumann of the Technical University of Munich worked to discover whether or not the

Earth has a frequency. His assumption about the existence of this frequency came from his understanding that when a sphere exists inside of another sphere, an electrical tension is created. Since the negatively charged Earth exists inside the positively charged ionosphere, there had to be tension between the two, giving the Earth a specific frequency. Through a series of calculations he came up with a frequency of 10hz that he believed to be the pulse of the Earth.

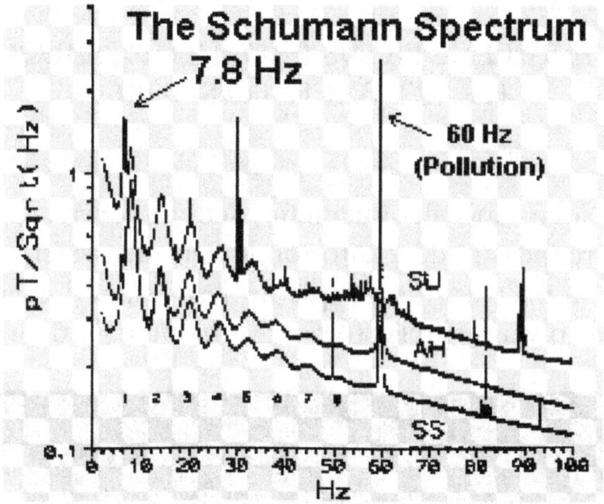

In 1954 Schumann teamed up with Herbert König, and confirmed that the resonance of the Earth maintained a frequency of 7.83 Hz. Since then the Schumann Resonance has been the accepted term to describe and measure the pulse or heartbeat of the earth.

Humanity's heart rhythms are synchronized on a global scale with the earth's energetic systems as well as every biological system on the planet, all of which are exposed to invisible magnetic fields of all kinds at every range of frequencies. These fields can affect every cell and circuit to a greater or lesser degree. A number of physiological rhythms have also been shown to be synchronized with solar and geomagnetic activity.

Human regulatory systems are designed to adapt to daily and seasonal climatic and geomagnetic variations. Sharp changes in solar and geomagnetic activity and geomagnetic storms can stress these regulatory systems resulting in alterations in melatonin/serotonin balance, blood pressure, immune system, reproductive, cardiac, and neurological processes. Disturbed geomagnetic activity is associated with the intensification of diseases, increases in myocardial infarctions, death, changes in blood flow, aggregation, and coagulation, increased blood pressure, cardiac arrhythmias, and seizures in epileptics.

During periods of increased solar activity every 10.5 to 11 years, the sun emits increased ultraviolet energy and solar radio flux. These solar and magnetic influences affect a wide range of human health and behavioral processes with the cardiovascular and nervous systems being affected the most.

In one study, ten healthy individuals from the ages of 34 to 65 years old had their Heart Rate Variability recorded for 31 consecutive days. Overall, the study suggested that daily autonomic nervous system activity not only responds to changes in solar and geomagnetic activity, but is synchronized with the time-varying magnetic fields associated with geomagnetic field-line resonances and Schumann resonances. A likely explanation for how solar and geomagnetic fields can influence human nervous system activity is through a resonant coupling between our nervous systems and geomagnetic frequencies called Alfvén waves, ultra-low frequency standing waves in the earth-ionosphere resonant cavity at Schumann resonances that overlap with physiological rhythms.

Another study published in the journal *Nature Communications* suggested that a protein in the human retina, when placed into fruit flies, has the ability to detect magnetic fields, serving as a magneto sensor, but it's unknown whether humans use it in this way.

In line with Sheldrake's hypothesis of morphic resonance within morphogenetic fields, researchers *have concluded that:*

- The Earth's magnetic fields are a carrier of biologically relevant information that connects all living systems
- Every person affects this global information field. Large numbers of people creating heart-coherent states of love, appreciation, care, and compassion can generate a more coherent field environment that benefits others and helps offset the current planetary discord and incoherence
- There is a feedback loop between human beings and Earth's energetic/magnetic systems
- Earth has several sources of magnetic fields that affect us. Two of them are the geomagnetic field that emanates from the core of the Earth, and the fields that exist between Earth and the ionosphere. These fields surround the planet and act as protective shields blocking out the harmful effects of solar radiation, cosmic rays, sand, and other forms of space weather. Without these fields, ice as we know it could not exist on Earth. They are part of the dynamic ecosystem of our planet.

The Earth and its ionosphere generate frequencies that range from 0.01 hertz to 300 hertz, some of which are in the same frequency range as the one in our brain, cardiovascular system, and autonomic nervous system. This fact explains how fluctuations in the Earth's and Sun's magnetic fields can influence us. Changes in these fields also affect our brain waves, heart rhythms, memory, athletic performance, and overall health.

Changes in the Earth's fields from extreme solar activity have been linked to some of humanity's greatest creations of art, as well as some of its most tragic events. We know how these fields affect us, but what about how we affect these fields? Scientists believe that because brainwave and heart rhythm frequencies overlap the Earth's field resonance, we are not just receivers of biologically relevant information, but also feed information into the global field creating a feedback loop with the Earth's magnetic fields. Research indicates that human emotion and consciousness encode information into the geomagnetic field and this encoded information is distributed globally. The Earth's magnetic fields act as carrier waves for this information which influences all living systems and the collective consciousness.

Neurocardiologist research reinforces the fact that we have an energetic relationship with every living system around us with processes that can affect our nervous system, health, emotions, and mood. The best way to work with this information and incorporate it into our lives is through self-observation, another term for awareness.

Scientists studying the intimate relationship between the brain and the heart have discovered that the heart has many parallels to the brain. It has about 40,000 specialized neurons that can sense, feel, learn, and remember, forming a sophisticated communication network within itself. This intricate network of neurons, nerves, neurotransmitters, proteins, and more found in the heart are identical to those found in the brain.

Science is only beginning to catch up with traditional beliefs when explaining experiences like intuition, precognition, self-healing, empathy, and abilities like telepathy and clairvoyance. This is especially clear when we examine the principles offered in some of the most cherished spiritual traditions, among them Ayurveda, Daoism, Sufism, Traditional Chinese Medicine, Mayan, and many others. Historical teachings demonstrate a deeper understanding of the heart's intelligence at the level of having direct influence upon our mind, personality, daily decisions, intuition, and our ability to make moral choices, and can be found in almost all traditions worldwide.

In the groundbreaking book *The HeartMath Solution*, authors Lew Childre and Howard Martin distill over 30 years of research on the complex role the heart plays in health, happiness, well-being, and longevity that reveals the heart to be a command center of formidable intelligence and intuitive knowledge that is tied into every system and every cell of our bodies. Here are their key discoveries.

8 Facts About The Heart's Intelligence

1. The heart contains its own independent nervous system, comprised of more than 40,000 neurons.

2. The heart's electromagnetic field is the most powerful field generated by your body. It is projected throughout the body, and radiates several feet outside of you.

3. Your heartbeat, rather than being a simple "lub-dub, lub-dub," is a tightly compressed information stream, broadcasting its complex messages to every cell throughout the entire body.

4. The heart provides you with emotional and intuitive guidance to help you direct your life in the positive direction you want it to go in.

5. Core heart feelings of love, appreciation, and compassion down-regulate the activity of the sympathetic nervous system, modulating the fight or flight response and increasing the activity of the parasympathetic nervous system to facilitate rest and digestion.

6. Your heart rhythms are mirrored in your emotional states. Negative emotions like fear, anger, and hostility create disordered and irregular heart rate variability, which is the healthy variation in the time interval between heartbeats. Positive emotions create improved order in the heart's rhythms, bringing greater emotional stability.

7. The heart is the body's master oscillator. Its rhythm pulls all the body's systems into a state of entrainment or synchronization.

8. Positive emotional states have a balancing effect on the nervous system by strengthening immunity, enhancing hormonal function, and improving brain function.

INFINITE PERMUTATIONS OF THE IMPLICATE ORDER AND THE PRIMARY SCISSION

Synchronicities refer to a subjective experience that coincides between events in the mind and the outside world. These events can be causally unrelated to each other and still have other unknown connections. Any connection implies resonance, and resonances are self-mirroring reflections within reflections, originating from an infinite Source of holograms nested within holograms.

This psychic mirroring is thought to be the basis for telepathy, resonance, animal communication, and any other method of tuning in that harmonizes with the energy of the Source in the same way that a radio receiver picks up a transmitted signal. This tuned in reflection also works the same way the secondary beam of a hologram does when it rejoins the Source to create an interference pattern on a recording medium.

This ubiquitous mirroring paradigm underlies ancient beliefs of telepathy that connect the human heart and its resonance with the Source beam of the sun when it matches its frequency, bringing resonance, telepathy, and open-heartedness.

The infinite point of origin from the Source constitutes the primary laser beam of a hologram before the split into duality and the subsequent reflection in the form of an interference pattern that forms the foundation of three dimensional perspective and perception.

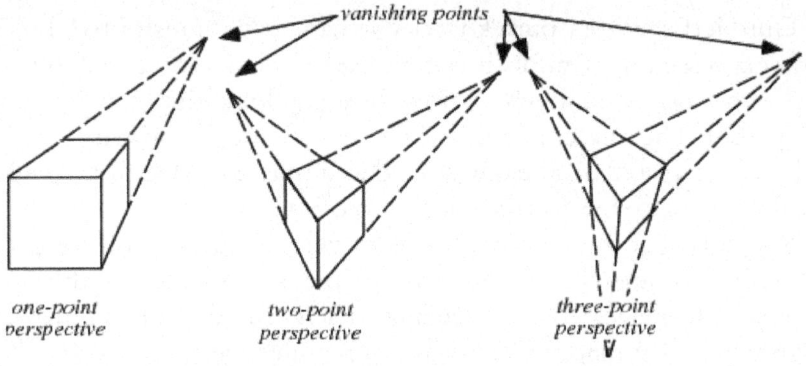

Points of Origin Represented by Vanishing Points

In the three dimensional reality that we inhabit, everything that evolves must have a beginning where all that it manifests comes into being. Like the wave-particle phenomenon, this nexus is the point of origin where the intangible becomes tangible, energy becomes mass in the form of matter, and the first duality that defines existence comes into play. Esoterically, this primary scission is represented by the vesica pisces and the infinity point that it represents is the basis for our dualistic three dimensional world view that is played out endlessly in our perceptions.

Spirit-matter, subject-object, positive-negative, yin-yang, male-female, day-night, sleeping-waking, wave-particle, and life-death are a small sampling of the dual archetype that provides the basis for life. It manifests in the inhale-exhale of our lungs, the beating of our hearts, the ticking of our clocks, our music in the primal beats of a drum, and the notes which would not be differentiated without the spaces between them.

Musical Note Duration for 4/4 Quarter Notes

Gurdjieff said that pure knowledge cannot be transmitted, but by being expressed in symbols it is covered by them as by a veil, but for those who desire and who know how to look this veil becomes transparent. The movement of the circle to create the primary scission of the vesica pisces can be envisioned as a drawing aside of the veil to reveal the expansive revelation it has to offer.

Regardless of the multiple forms of verbal expression throughout the world, geometry is solid, observable self-apparent truth in a universal language that articulates itself in dynamic unfolding progressions that map the pathways of manifesting reality.

Dimensionality is dictated by geometric expansion and contraction, whether opening itself out into the infinite holographic macrocosm or moving down through the equally infinite nested microcosm.

This primary movement from zero point to a line is that same first movement of the cosmic breath upon the waters of creation where not only is One created from zero, but the energy that created that state is evident within the two new polarities. When these polarities are transcended the paradox of their opposites is overcome, opening the progression up to more expanded levels of expression. One and zero in the natural world are also the basis for human logic that dictates not only the form and function of our world wide web of computers, but our entire monetary and business network as well.

What used to be backed by gold, one of the densest elements, is now many times removed and diluted from its original physical representation as promissory notes, to a series of ones and zeros representing a changing state of energy, one or zero, on or off, yes or no.

This endless duality permeates our existence with the beating of our hearts, the in and out of our breaths, the two sides of our bilateral brains, the rising and setting of the sun and moon, and in the interplay of masculine and feminine energies that make life possible. Shamans honor this most primal pulse of life and being through the beats of their drum, which is considered and extension of their heart.

The initial singular movement of a point into a line begins the one-dimensional angular progression of dimensionality from the zero point of manifestation out through the three dimensions of length, width, and height, which define our consensual perception of reality. The straight line angled expressions that we see in the two dimensions of squares and rectangles, and the three dimensions of cubes and boxes represent the Cosmic Masculine.

If we begin at the zero point of origin and extend out equally from its central location in every direction on the same plane that we extended the line on, we create a two dimensional circle, and if we extend the point out equally in every possible direction we have a three dimensional sphere that represents the Cosmic Feminine.

Ancient cultural examples of this prehistoric truth are evident on the island of Amantani situated on Lake Titicaca in Peru, close to the Bolivian border. Amantani is said to be a mystical power place because of the ancient temples located there. The island consists of 2 mountains with an ancient temple on each peak; the mother earth temple of Pachamama, dedicated to the feminine aspect of the universe, and the father earth temple of Pachatata, dedicated to the masculine aspect.

The masculine temple of Pachatata is very distinct sitting right on its peak in a square form, while the temple of Pachamama has a round shape, and the mountain it is located is more diverse as well as being wider and more open.

Both temples have very different energies. At Pachatata many people experience a sense of clarity, centeredness, inner strength and report a connection with their higher selves. At Pachamama they experience a deeper sense of beauty, love, nourishment, and connection with the earth.

Other deeply embedded masculine and feminine qualities follow these circular and angular geometric indicators of gender in both the human and animal kingdoms, further reinforcing this observation. When a man concentrates on a particular task he remains quiet with a single minded focus to the exclusion of all else, reinforcing the point to line directness, while women multitask, and "gossip" with each other.

These habits are inherent in human and animal behaviors, particularly mammals. A male hunter must be quiet, direct, and often solitary to be successful, while women, out of biological necessity must not only keep the nest, they must be aware of everything that goes on around them and any other nest keepers in the vicinity in order to survive. Aside from keeping a watchful eye over their offspring, their multitasking and communicative chatter is an essential skill for the gathering part of the hunter gatherer paradigm.

On another note, the two basic fears that humans have are divided between the sexes. Women, who are vulnerable because of the position and dependence that child bearing puts upon them fear abandonment, while men who are driven to be solitary and independent fear entrapment.

The sacred feminine circle carries high value in tribal societies. Structures are built round, tribal councils are held in circles, and medicine wheels are circular, all honoring the sun, moon, and the life giving circles evident in the seasons, orbits, and endlessly repeating cycles of nature.

The *Emerald Tablet* written by Hermes Trismegistus contains the secret of transmutation of primordial substance. The tablet text contains the famous lines "As above, so below" and goes on to explain creation in the context of duality. Understanding the duality of the masculine and the feminine is the key. Much of this wisdom has been lost through the ages resulting in grand misconceptions of what actually constitutes the masculine and feminine in their archetypal forms. Understanding these forces is more important now than ever, for it is the misunderstanding and the resultant extreme polarization of these forces that threatens life on earth. Mastery of both of these forces and the synthesis of the two creates growth, advancement, and value fulfillment.

Fundamentally, the masculine and the feminine are *not* two tangible separate things. They both exist on the same continuum and are very real forces having different directions while being inexplicably linked. One cannot exist without the other in any meaningful way, and for that matter no polarity whether considered masculine, feminine, positive, negative, light or dark can exist without its opposite.

Categorically the feminine and masculine can best be understood as forces or tendencies in the following ways:

Masculine — Feminine

Projective — Receptive
Expansive — Contractive
Heating — Cooling
Sympathetic — Parasympathetic
Intellectual — Emotional
Outward — Inward
Brighter — Darker
Yang — Yin
Magnetic — Dielectric
Square — Circle

The sun is a blazing fire that radiates it's energy in all directions making it the archetypal masculine in almost every way making "Father Sun" an apt description. The earth on the other hand is archetypal feminine, referred to as "Mother Earth", mostly dark blue water and a black soil planet that receives the bright masculine rays from the sun, making the sun projective and the earth receptive. The sun above and the earth below create the polarity where life and creation occurs in-between.

In the Taoist system yang is white symbolizing the projective light of the sun and yin is black symbolizing the receptive earth.

Yin and Yang Symbol

It is interesting to note that the color green is in the middle of the visible spectrum. Green is the color of life on earth and represents the meeting point between the red fire of the masculine and the cool nurturing blue of the feminine. Plants nourish themselves from water and soil in the feminine ground and send their branches and leaves skyward for nourishment from the masculine sun. Trees and how they function are highly symbolic of these primordial forces, being the creative point in the delicate balance between them.

The process of clouds condensing to cooler water and coming down from the sky and nourishing/replenishing life on earth is the feminine act of cooling, soothing, and rejuvenating.

In the human body arteries carry blood which is red (masculine) when oxygenated and the veins are purple/blue (feminine) and are depleted of the oxygen, yet receptive to it in the lungs for replenishment.

The Taoist system explains the story of creation from primordial source Wu-Chi which means "without ridgepole", and is considered an infinite boundless state of no differentiation. The dot in the middle is the point of yin that creation emerges from. The forces of yin then contract the yang to form a state of rotating force-balance or Tai Chi which means "great pole." Yang is a universal expansionary force of continuous rapid replication and yin is a contractive force that contracts rotationally in order to create reality.

Creation is an emergent property of the interplay between masculine and feminine and the story of creation is fundamentally a story of movement from non-duality/polarity to duality/polarity.

The Biblical parallel to the Taoist system of the creation of masculine and feminine is the story of how feminine Eve was created from a rib from the side of masculine Adam, and also how light, (yang) was the first creation, then all forms of matter (yin) and life were created after.

The cosmological analogy is in the early part of the theoretical "big bang" where the universe existed in a super hot state of expanding plasma, a masculine process, then through a feminine process in the formation of matter, planets, and stars that came through a process of coalescing and cooling.

Hyper-dimensionally the expansionary and contractory nature of the holographic universe goes out to infinity both inward and outward.

There is no beginning or end, and no first or last.

GETTING TO THE POINT

The birth of polarity in its archetypal form is the most fundamental necessity for existence. Without it no frameworks can be created. The feminine represents detailed refinement and a move toward ever increasing complexity and novelty in a higher order. The permutations of duality are endless, but to understand its origins we need to get to the point, where it all begins.

.

A point has no size and can only be imagined. It fixes a location in space and has a dimension equal to zero. If you take a point and move it from its original location to another location, this moving point creates a line which constitutes the first dimension known as length.

This movement is a manifestation of energy, which is the essence of consciousness — the proverbial wind on the water, the first movement of creation.

When we move a line which has no thickness from its original location to another location, it becomes a plane which has two dimensions, length and width.

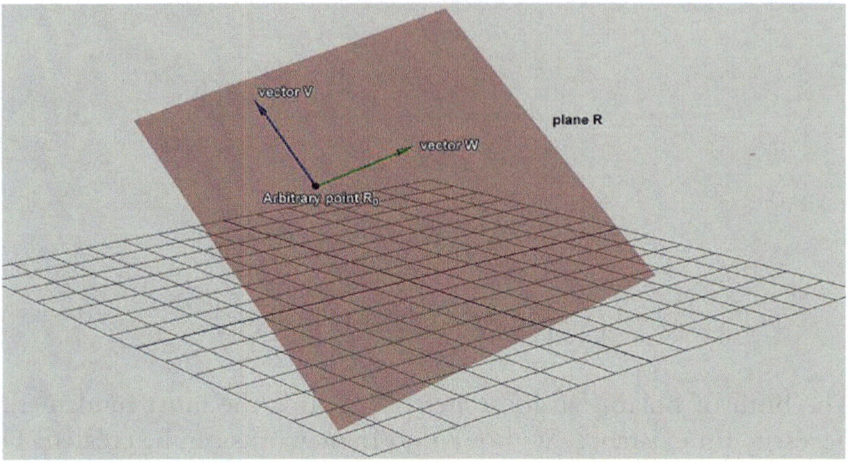

When we move a plane, it leaves two dimensions and creates a solid body with three dimensions, length, width, and height. If each of these dimensional shifts cover the same distance, the result is a cube, the basic representation of three dimensional reality represented by 2^2, known as two to the second power, or two cubed.

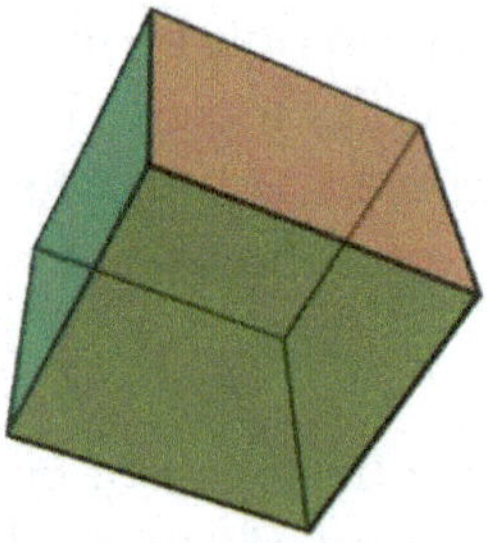

When you move a three dimensional body in space, it remains a three dimensional body and does not leave the third dimension.

If you imagine yourself without any dimensions you would be formless. The point represents zero dimension. The first dimension indicating energetic movement is a line, the second dimension is a surface, and the third dimension is a solid body. If you imagine yourself as a being who can only move along the first dimension of a straight line, all you would see is points, not your own dimensionality, because when we attempt to draw anything within a one dimensional line, points are the only option.

A two-dimensional being moving through a plane would encounter lines and thus distinguish one-dimensional beings, and a three-dimensional being within a cube would encounter planes and perceive two dimensional beings.

Human beings perceive in three-dimensions.

By the logic of this progression it becomes apparent that the perception of each dimension necessitates being one dimension above it, so human beings who can delineate external beings in three dimensions and manipulate three-dimensional spaces have to be four dimensional beings. In the same way that being within a cube can only perceive two dimensions and not its own third dimension, human beings cannot perceive the fourth dimension.

The movement of the point to make the line, the line to the plane, and the plane to the solid all take time, making time synonymous with movement. We perceive the energetic movement of our three dimensional bodies through space as time. How long does it take to get from point A to point B? Our whole concept of time and our calendar is based on physical movement, specifically by counting the number of identical cyclical movements that a particular time measurement corresponds to.

It takes three hundred and sixty five and a quarter days for the earth to circle the sun in a solar year, twenty eight days for the moon to circle the earth, a day of twenty four hours for the earth to complete a rotation on its axis, sixty minutes to make an hour, and sixty seconds to make a minute.

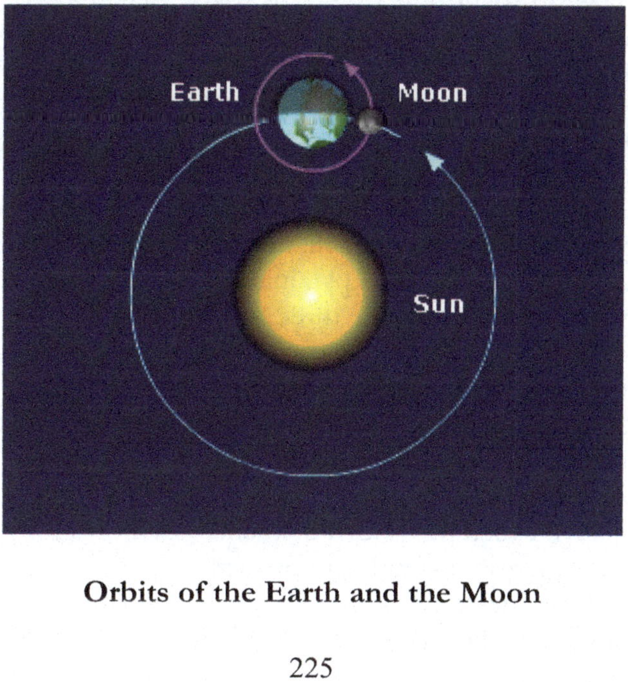

Orbits of the Earth and the Moon

Our materialistic world view has us locked in to time perception based on the movement of physical bodies through space, but what of the source of the expansion and its point of origin that make up the invisible half of the formula?

This formless half of the formula comes from inside of us where the seat of consciousness resides, perched between subject and object in an interference pattern or infinity zone where the evolution of consciousness takes place. Following the geometry and logic of these progressions, a shift in our perception would bring a shift in how we live in and perceive the world from our present fourth dimensional existence that perceives a three dimensional world, to fifth dimensional existence where we perceive the fourth dimension, transcending time and duality.

Quantum physics have proven that matter is more empty space than solidity. In terms of Einstein's $E=MC^2$, matter is another form of energy, and energy is the essence of consciousness. The fundamental nature of this paradox tells us that positive and negative are extremes. It is the energy between them that matters. That is where we exist within ourselves; at the threshold of subject and object, in the center, mid-point of our duality. When we transcend the poles, we become centered and rise above the opposition of forces to embrace the combined energy that they manifest in this unity of opposites.

When the mid-point rises above the baseline, its transcendence creates a triad, trinity, or triangle. The three dimensional representation of a triangle with four sides creates a pyramid, one of the greatest enigmas known to mankind. The square base of a pyramid represents four dimensions and the four directions of the physical world and rises up pointing toward the heavens in a diminishing apex that represents an infinity point of origin where matter ends and spirit begins.

Similar to the example of a pyramid, the laws of geometry are self-evident when moving from three dimensional space to fourth dimensional time through the study of projective geometry.

Euclidean geometry describes shapes as they are. Their properties of length, angles, and parallelism are unchanged by rigid motions.

Projective geometry describes objects as they appear. Lengths, angles, and parallelism become distorted when we look at objects from this perspective which provides a mathematical model for how images of the three dimensional world are formed.

Projective geometry is the branch of mathematics that deals with the relationships between geometric figures and the images, or mappings that come from projecting them onto another surface.

Common examples are the shadows cast by opaque objects, motion pictures displayed on a screen, and the creation of a hologram.

As shown below, the eye is connected to points on the landscape of the horizontal reality plane, RP, by sight lines. The intersection of these lines with the vertical picture plane, PP generates the drawing, projecting the reality plane onto the picture plane, hence the name projective geometry.

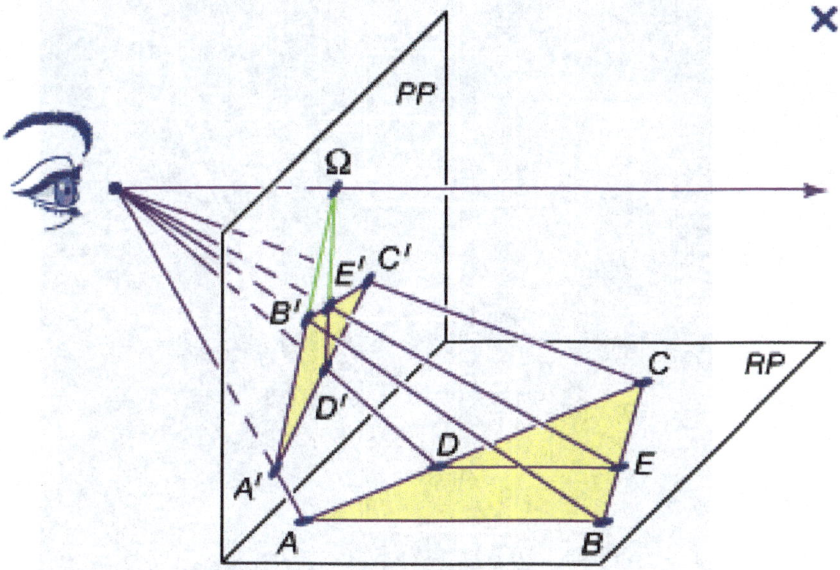

Projective Geometry

Counter space is the space where forces not amenable to ordinary measurement work. Instead of having its ideal elements in a plane at infinity it has them in a "point at infinity". These are lines and planes rather than lines and points as in ordinary space, and are known as *counter space infinity* and only appears for a different, peripheral kind of consciousness that experiences such a point as infinite inwardness in contrast to normal consciousness that experiences infinite outwardness.

In counter space terms, the Infinity Zone is known as a linkage; an element that belongs to both Euclidian and counter space at once. A point or plane like this can be considered as another manifestation of nonlocality and the holographic implicate order. If a cube is linked to both spaces at once and is moved upward away from the inner infinitude, it will obey the metrics of both spaces. The diagram below shows what happens as it moves. The smaller version obeys space and stays the same size and shape in space while the bigger version obeys the counter space metric.

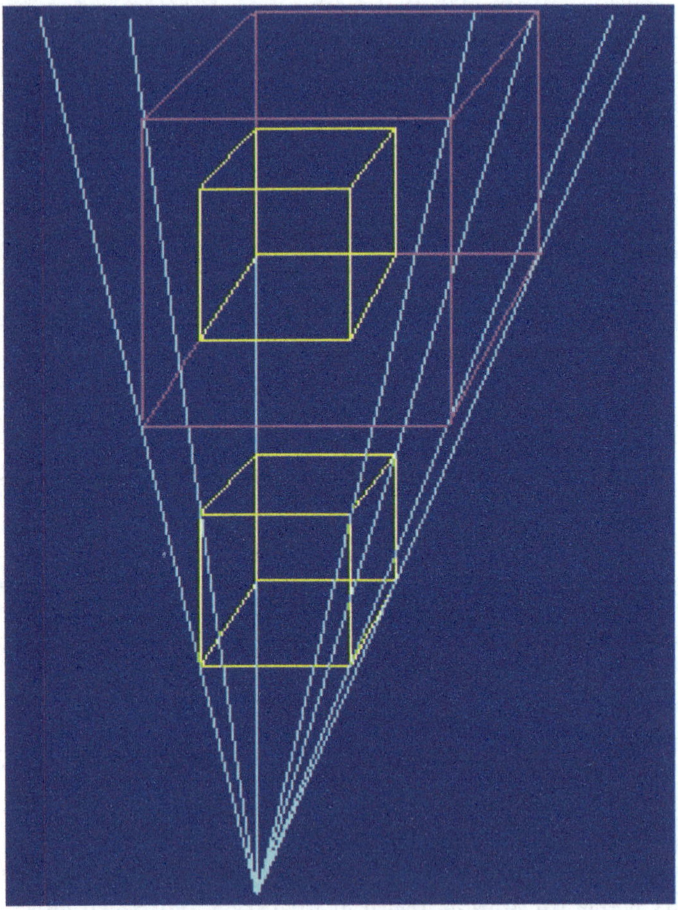

The counter space or inner infinity is shown as a point at the bottom. Lines drawn from it through the vertices of the cube stay on these lines, obeying its metric properties, illustrated by the smaller cube, while the smaller cube stays the same spatially. This seems natural with our ordinary consciousness, but for counter space consciousness the other is the most natural and the projected cube appears to be getting bigger.

The geometric difference between the two cubes is referred to as *strain,* analogous to the use of that term in engineering where it is the percentage deformation in size, like when a rubber band is stretched. The rubber band responds to the strain by exerting a force referred to as *stress.*

The central principles here are:

- Objects may be linked to both spaces at once.
- When they are, strain arises when they move because the metrics are conflicting.
- Stress arises as a result of the strain.

Stress is not a geometric concept so we move from geometry to physics when we consider it. The major stress free movement or transformation is rotation about an axis through the counter space infinity represented by sideways figure **8** - **∞**. This focal "power point" axis that is the core of stress free movement and transformation is the infinity zone, the zero point threshold where we transcend polarities and become centered.

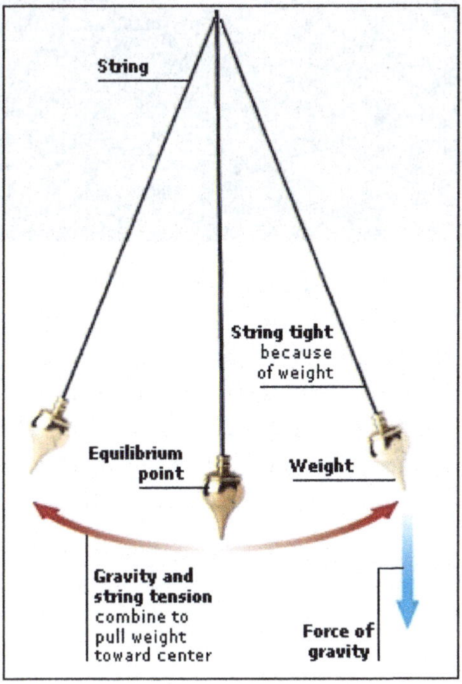

As an example, when a pendulum swings down it moves at its greatest speed when it passes the middle point and goes slower and slower as it rises on the other side until it reverses its movement. The greatest amount of energy is at the center, as opposed to the extremes where the power is the weakest. The greatest power in our solar system lies in the sun that resides at its center. If you look at a terrestrial energy manifestation like a hurricane, one of the most powerful forces of nature, it revolves around the eye of the storm, which is its calmest place; its center of power, the heart that generates a Fibonacci spiral that originates in phi, and the root of the word inspire.

Solar System

In our dualistic world view we focus on the extremes, falsely believing them to be the places of maximum power, but the greatest power resides in understanding that it is not the polarities that deserve our attention, it is the energy at the center in the "eye of the storm", the zero point of the infinity zone where nothing happens.

EVERYTHING IS NOTHING AND NOTHING IS EVERYTHING

Nothing is a concept denoting the absence of something, and is associated with *nothingness,* the state of being nothing, the state of nonexistence of anything, or the property of having nothing. This is how "non-believing" atheists characterize any conception of God, yet in the paradox of their denial they are unequivocally announcing that *they themselves are God* as they have made the ultimate judgment call, claiming their throne at the center of *their* universe.

From this perspective, if they ceased to exist then nothing else would exist, which makes them a creator God who rules over everything that falls under their range of thought and perception, and denies existence to any other higher intelligence that might exist "outside" of them.

The idea that God created the universe out of nothing (*creatio ex nihilo*) became central to Judaism, Christianity, and Islam. With the exception of the metaphorical wind moving over the water that hinted at this in the first sentence of the Hebrew Bible, this concept is not found in Genesis, nor in the entire Hebrew Bible. The authors of Genesis 1, writing around 500–400 BCE were not concerned with the origins of matter, the material that God formed into the habitable cosmos, but with the fixing of destinies. This was the situation in the early second century CE, although early Christian scholars began to see tension between the idea of world-formation and the omnipotence of God. By the beginning of the third century this tension was resolved, world-formation was overcome, and creation *ex nihilo* became a fundamental tenet of Christian theology.

Ex nihilo is a Latin phrase meaning "out of nothing" and often appears in conjunction with the concept of creation, as in *creatio ex nihilo*, meaning "creation out of nothing", chiefly in philosophical or theological contexts, but it also occurs in other fields.

In theology the phrase *creatio ex nihilo* contrasts with *creatio ex materia* which is creation out of some pre-existent, eternal matter, and with *creatio ex deo*, creation out of the being of God. *Creatio continua* is the ongoing divine creation.

The phrase *ex nihilo* also appears in the classical philosophical formulation *ex nihilo nihil fit*, which means "Out of nothing comes nothing."

When used outside of religious or metaphysical contexts *ex nihilo* refers to something coming from nothing. For example, in a conversation, one might raise a topic *"ex nihilo"* if it bears no relation to the previous topic of discussion.

One of the most respected elemental spirits among tribal cultures is the Wind, an invisible force capable of unfathomable power that has no substance, yet is defined as not only a direct link to spirit, but *spirit itself* manifest as formless, invisible, energy. We have no perception of it when it is at rest and we perceive and characterize it as *nothing* as it registers nothing with any of our senses, making it a perceptual enigma, yet it is the most pervasive presence we know.

We cannot speak, act, or think without its participation and this unseen enigma holds the mystery that enables life to live and unites our breathing bodies to the world around us, and to the interior life of all that we perceive in the open field of this living presence.

In shamanic thought the air is the soul of the visible landscape that constitutes the secret realm where all beings draw their nourishment from, yet it is that most intimate absence felt and characterized as *nothing.*

Nothing is more common to the diverse indigenous cultures of the earth than a recognition of the air, the wind, and the breath as aspects of a singularly sacred power. By virtue of its pervading presence, its invisibility, and its influence on all manner of visible phenomena, the air is the archetype of all that is ineffable, unknowable, yet undeniably real and efficacious. Many indigenous people construe awareness not as a power that resides inside their heads, but as a quality that they themselves are inside of.

All of this can also be defined as a void, meaning a completely empty space. The contradictory fact that void is defined as a noun, which by definition is a person, place, or thing, means that we have defined nothing as something, which is nothing. This puzzling contradiction also encompasses the term point which has no dimension, meaning zero, and spirit which is formless and invisible.

They all mean the same thing.

Nothing.

Zero-point or ground state energy is considered the lowest possible energy that a quantum mechanical system can have. Unlike classical mechanics quantum systems fluctuate in their lowest energy state due to the Heisenberg uncertainty principle. According to modern physics the universe can be thought of not as isolated particles but continuous fluctuating fields: matter fields whose quanta are fermions like leptons and quarks, and force fields, whose quanta are bosons characterized as photons and gluons. All these fields have zero-point energy and these fluctuating zero-point fields lead to a reintroduction of an aether, the postulated medium for the propagation of light.in physics.

Physics lacks a full theoretical model for understanding zero-point energy, in particular the discrepancy between theorized and observed vacuum energy which has become a source of major contention. Physicists Richard Feynman and John Wheeler calculated the zero-point radiation of the vacuum to be an order of magnitude greater than nuclear energy, with one teacup containing enough energy to boil all the world's oceans, yet according to Einstein's theory of general relativity any such energy would gravitate. The experimental evidence from both the expansion of the universe, dark energy, and the Casimir effect, which is defined as a small attractive force that acts between two close parallel uncharged conducting plates due to quantum vacuum fluctuations of the electromagnetic field, show any such energy to be exceptionally weak.

If supersymmetry is valid at all, it is a broken symmetry, only true at very high energies, but no one has been able to show a theory where zero-point cancellations occur in the low energy universe we observe today. This discrepancy is known as the cosmological constant problem, or vacuum catastrophe. This disagreement between the observed values of vacuum energy density, which is the small value of the cosmological constant, and the theoretical large value of zero-point energy suggested by quantum field theory is one of the greatest unsolved mysteries in physics.

In mathematics, "**0**" is the integer preceding **1** and it is an even number because it is divisible by **2** with no remainder. **0** is neither positive nor negative. By most definitions it is the only natural number not to be positive. Zero is a number which quantifies a count or an amount of null size. In most cultures, **0** was identified before the idea of negative things where quantities less than zero were accepted.

The number **0** is the smallest non-negative integer. The natural number following **0** is **1** and no natural number precedes **0**. The number **0** may or may not be considered a natural number, but it is a whole number and hence a rational number and a real number as well as an algebraic number and a complex number.

Being neither positive or negative **0** is usually displayed as the central number in a number line, and it is not a prime number or a composite number. It cannot be prime because it has an infinite number of factors, and it cannot be composite because it cannot be expressed as a product of prime numbers. Zero has to always be one of the factors, but it is even as well as being a multiple of any other integer, rational, or a real number. A common paradox occurs with mathematical idealizations like point sources that describe physical phenomena at distant or global scales that break down at the point itself.

These paradoxes are sometimes seen as relating to Zeno's paradoxes which deal with the physical manifestations of mathematical properties of continuity, infinitesimals, as well as infinities associated with space and time. As an example, the electric field associated with a point charge is infinite at the location of the point charge. A consequence of this paradox is that the electric field of a point-charge can only be described in a limiting sense by a carefully constructed Dirac delta function. This mathematically inelegant but physically useful concept allows for the efficient calculation of the associated physical conditions while conveniently sidestepping the philosophical issue of what actually occurs at the infinitesimally-defined point: a question that physics has been unable to answer.

In Euclidean geometry a point is a primitive notion that the geometry is built upon. Being a primitive notion means that a point cannot be defined in terms of previously defined objects. A point is defined only by properties called axioms that it has to satisfy. Geometric points do not have any length, area, volume, or any other dimensional attribute. A common interpretation is that the concept of a point is meant to capture the notion of a unique location in Euclidean space.

Points considered within the framework of Euclidean geometry are one of the most fundamental objects. Euclid originally defined the point as "that which has no part."

There are several inequivalent definitions of dimension in mathematics and in all of the common definitions a point has no dimension.

Like the Wind, the Air, and the Nothing that gives us existence, there are no physical proofs of electricity, magnetism, or gravity, yet the tangible proof of their existence is inherent in the invisible world around us that provides the medium that they travel through. They exist as manifestations and forms of energy, and though we cannot see their forms we can see and harness their effects.

It is interesting to note that in terms of "invisible" energy we cannot see emotions, especially in others, but we can perceive their effects inside ourselves. The emotional perception and empathy we experience from being tuned in to emotions and being heart centered indicates where spirit or essence as Source meets the world.

THE INFINITE POWER OF ZERO IN FORM AND MOTION

The mathematical symbol that represents three dimensional form and fourth dimensional motion is a sideways figure **8** - ∞. The central point of this infinity symbol where transition occurs represents the invisible center of the infinity zone where the greatest power lies within the zero point of nothing.

It is no accident that tremendous power has been attributed to the number **8** which has great meaning and significance throughout history the world over. It is considered the most powerful of all numbers in numerology.

The number **8** is the base figure of the octal number system and the axis around which the number systems evolved across the globe. The most commonly used decimal system can be chronologically traced back to the octal system, adding more substance to the belief that the number **8** is the number of regeneration and new life; in other words, the numerical representation of Source.

The properties, qualities, and beliefs around the number **8** are seemingly infinite themselves. Its multitude of attributes have been culled to give examples from the many realms that it influences.

In Mathematics:

- **8** is **a** composite number, its proper divisors being 1**,** 2, and **4.** It is twice **4** or four times **2**. Eight is a power of two, being 2^3.
- The Prime Factors of **8=2x2x2.**
- **8** is the base of the octal number system mostly used with computers. In octal, one digit represents 3 bits. In modern computers, a byte is a grouping of eight bits, also called an octet.

- **8** is the largest cube in the Fibonacci Sequence, being **3** plus **5**. The next Fibonacci number is 13. 8 is the only positive Fibonacci number, aside from 1, that is a perfect cube.
- The maximal number of regions into which a plane can be divided by 3 circles = 8
- **8** is the maximal number of regions into which space can be divided by 3 Spheres.
- A polygon with eight sides is an octagon.
- Figurate numbers representing octagons, including eight are called octagonal numbers. A polyhedron with eight faces is an octahedron. A cuboctahedron has as faces six equal squares and eight equal regular triangles.
- A Truncated Tetrahedron is an Archimedean Solid with 8 faces, four triangles and four hexagons.
- The Cube is **a** Platonic Solid with 8 Vertices.
- Sphenic numbers which are positive integers that are the product of three distinct prime numbers always have exactly eight divisors.
- A figure **8** is the common name of a geometric shape, often used in the context of sports like skating. Figure-eight turns of a rope or cable around a cleat, pin, or bitt are used to belay something.

In Physics:

- In nuclear physics **8** is the second magic number.
- In particle physics the eightfold way is used to classify sub-atomic particles.

In Astronomy:

- As of 2006, in our Solar System, eight of the bodies orbiting the Sun are considered to be planets.

In Chemistry:

- The atomic number of oxygen.
- The number of allotropes of carbon.
- The most stable allotrope of a sulphur molecule is made of eight sulphur atoms arranged in a rhombic form.
- The maximum number of electrons that can occupy a valence shell.

In Biology:

- All spiders, and more generally all arachnids have eight legs. Orb-weaver spiders of the cosmopolitan family Areneidae have eight similar eyes.
- The octopus and its cephalopod relatives in genus Argonauta have eight arms as tentacles.

In General:

- Timothy Leary identified a hierarchy of eight levels of consciousness.
- In human adult dentition there are eight teeth in each quadrant. The eighth tooth is the so-called wisdom tooth.
- There are eight cervical nerves on each side in man and most mammals.
- In chess, each side has eight pawns and the board is made of 64 squares arranged in an eight by eight lattice. The eight queens puzzle is a challenge to arrange eight queens on the board so that none can capture any of the others.
- Eight is the third number that stays the same when written upside down.
- Scorpions have eight legs.
- According to Indian mythology, the Earth is supported on the backs of eight white elephants.
- Before the rise of Christianity, there were eight days in the Greek and Roman weeks.
- Many words beginning *oct-* are related to the number eight. An octopus has eight arms and an octet is a group of eight musicians.
- An octagon is a figure with eight sides and an octahedron has eight faces.

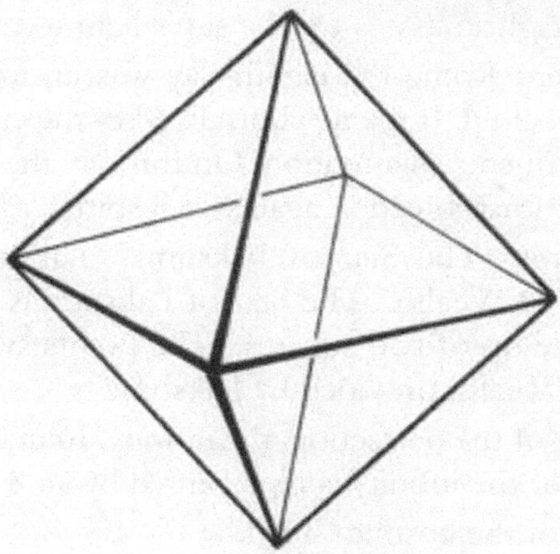

Eight Triangular Faces on an Octahedron

1	x	**8**	+	**1**	=	**9**
12	x	**8**	+	**2**	=	**98**
123	x	**8**	+	**3**	=	**987**
1 234	x	**8**	+	**4**	=	**9 876**
12 345	x	**8**	+	**5**	=	**98 765**
123 456	x	**8**	+	**6**	=	**987 654**
1 234 567	x	**8**	+	**7**	=	**9 876 543**
123 456 789 x **8** + **9** = **987 654 321**						

Mathematical Magic with the Number Eight

The number 8 also has the following symbolic attributes:

- Eight symbolizes the ability to make decisions.
- Eight symbolizes abundance and power.
- The Pythagoreans called the number eight "Ogdoad" and considered it the "little holy number".
- In China eight is homonym for prosperity. When pronounced it sounds much like the word "prosper".
- A double eight, as in **88,** is said to bring double joy.
- The eighth day of the Chinese New Year is the day for the annual gathering of all the gods in Heaven.
- In the Tarot, eight is the card for Justice or Strength.

- The stop sign has eight sides.
- It takes eight minutes for the sun's light to reach the earth.
- In Ancient Rome the eighth day was an important day to a newborn child. If the newborn lived to the eighth day the child was worth special attention. On this day the baby was rubbed with salt for protection against evil spirits.
- Hinduism - The Star of Lakshmi. Lakshmi is the Hindu goddess of Wealth. The Star of Lakshmi is the eight pointed star made up of two squares. The points symbolize the eight kinds of wealth provided by Lakshmi.
- Number of the perfection, the infinity. In mathematics the symbol of the infinity is represented by an **8** laid down.
- Symbol of the cosmic Christ.
- Number figuring the immutable eternity or the self-destruction. It represents also the final point of the manifestation.
- In China, **8** expresses the totality of the universe.
- Number of the balance and of the cosmic order, according to the Egyptians.
- Number expressing matter, it is also the symbol of the incarnation in matter which becomes itself creative and autonomous, governing its own laws.
- The number eight corresponds to the New Testament, according to Ambroise.
- It is the symbol of the new Life, the final Resurrection and the anticipated Resurrection that is the baptism.
- According to Clement of Alexandria, the Christ places under the sign of **8** the one he made to be born again.
- Represents the totality and the coherence of the creation in evolution. In China, it expresses the totality of the universe.
- Represents the earth, not in its surface but in its volume, since **8** is the first cubic number.
- The Pythagoreans made the number **8** the symbol of love and friendship, prudence, and thinking. They called it the Great "Tetrachtys".
- In Babylon, Egypt, and Arabia, it was the number of the duplication devoted to the sun, from where the solar disc is decorated with a cross of eight arms.

- The number 8 means multiplicity for the Japanese.
- A favorable number, associated to the prosperity.
- It is the number of the restful day, after the 7th day of the creation.
- The number eight governs the life of man: at 8 months baby teeth appear; at **8** years, he loses them; at 2 x 8 years it is puberty; and he becomes impotent at 8 x 8 years.
- Eight is the total number of Chakras of man, counting the seven in correlation with the physical body, plus an additional working in the etheric body. This eighth Chakra is known as "the Chakra of the Soul" or "the Star Chakra", located approximately **7** to 10 centimeters above the Crown Chakra.
- The eight great gods of the Vedas: Surya, Candra, Agni, Yama, Varuna, Indra, Vâyu and Kubera.
- The eight parts of the way that lead to nirvana, according to the Buddhist doctrines: the faith, the right judgement, the right language, a right and pure action, a right profession, the application of the spirit to all the precepts of the law, the right memory, and the right meditation.
- The Buddhists count eight symbols of long life of which one of them is the infinite node being rolled up and withdrawn on itself.
- There are eight degrees of Buddhist monks or "Aryas" and the highest is named "Arhat".
- According to a Buddhist legend, Buddha's ashes were separated in eight parts.
- The lotus is represented symbolically with 8 petals and according to yogis, it is on a lotus with 8 petals that the "Mêru" mount is erected which symbolically represents the center and the axis of the world.
- Yoga counts eight training courses: Yama, the restriction; Niyama, religious observances; Asana, the posture; Pranayama, the control of the breathing; Praty-ahara, the restriction of senses; Dharana, the concentration; Dhyana, the contemplation and Samadhi, the ecstasy.
- The eight "trigrams" of "Fou-Hi" Chinese philosopher.

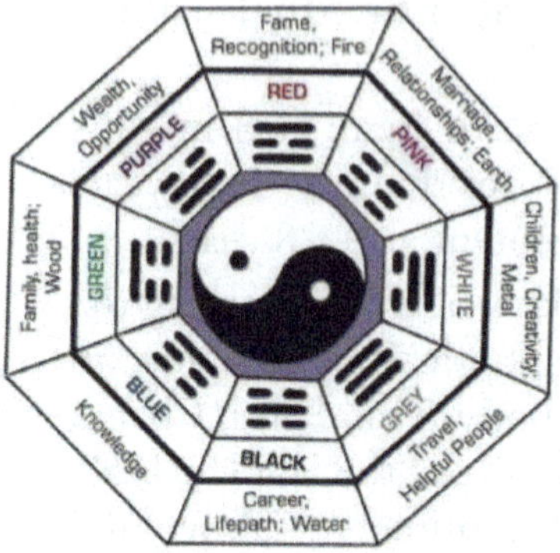

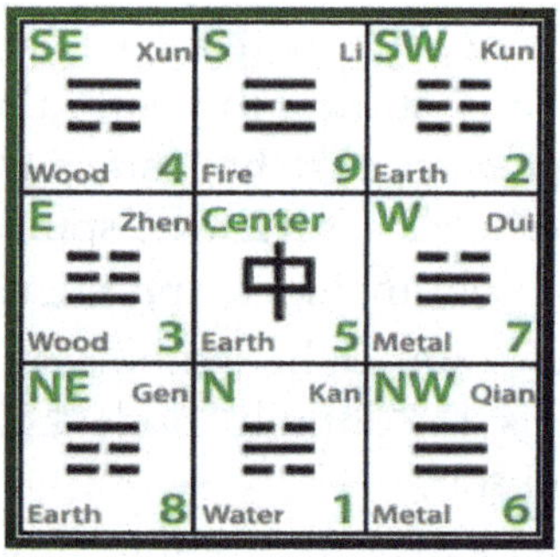

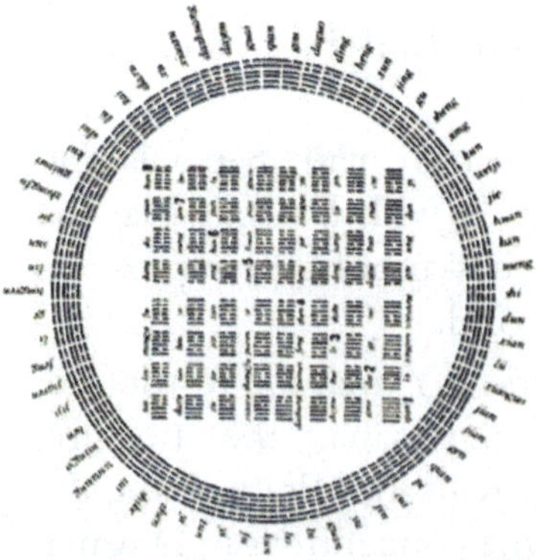

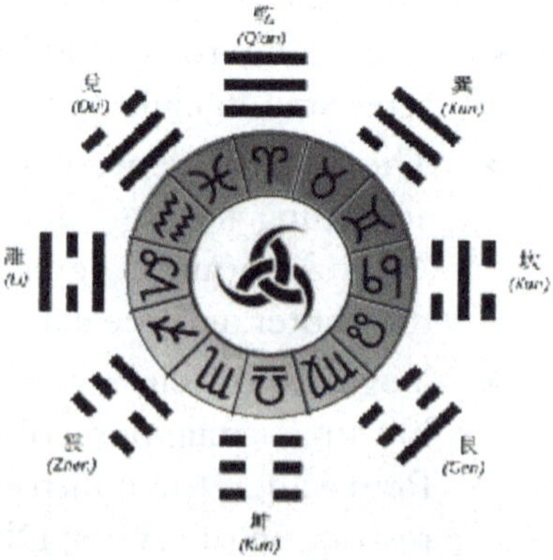

Chinese Trigrams

In Buddhism, the 8-spoked Dharmacakra represents the Noble Eightfold Path.

Division	Eightfold Path factors	Acquired factors
Wisdom (Sanskrit: *prajñā*, Pāli: *paññā*)	1. Right view	9. Superior right knowledge
	2. Right intention	10. Superior right liberation
Ethical conduct (Sanskrit: *śīla*, Pāli: *sīla*)	3. Right speech	
	4. Right action	
	5. Right livelihood	
Concentration (Sanskrit and Pāli: *samādhi*)	6. Right effort	
	7. Right mindfulness	
	8. Right concentration	

In Mahayana Buddhism, the branches of Buddha's Noble Eightfold Path are embodied by the Eight Great Bodhisattvas:

- Manjusri
- Vajrapani
- Avalokiteśvara
- Maitreya
- Ksitigarbha
- Nivaranavishkambhi
- Akasagarbha
- Samantabhadra

These are later associated with the Eight Consciousnesses according to the Yogacara school of thought: consciousness in the five senses, thought-consciousness, self-consciousness, and unconsciousness-consciousness referred to as alaya-vijñana. The irreversible state of enlightenment when a Bodhisattva goes on autopilot is the Eight Ground or *bhūmi*. Eight is an auspicious number for Buddhists. It has eight auspicious symbols, among them the jewel-encrusted parasol, the goldfish, always shown as a pair, the glyph of Pisces, the self-replenishing amphora, the white *kamala* lotus-flower, the white conch, the eternal Celtic-style infinitely looping knot, the banner of imperial victory, the eight-spoked wheel that guides the ship of state that symbolizes the Buddha's teaching. Similarly, Buddha's birthday falls on the **8**th day of the 4th month of the Chinese calendar.

In superstition and divination the number 8 as a sign of infinity appears as a lucky number.

Eight (八; accounting 捌; pinyin *bā*) is considered **a** lucky number in Chinese culture because it sounds like the word meaning to generate wealth (發(T) 发(S); Pinyin: *fā*).

Property with the number 8 is valued by the Chinese. A Hong Kong number plate with the number 8 sold for $640,000.

In the 2008 Games of the XXIX Olympiad**,** the official opening was on 08/08/08 at 8:08:08 p.m. local time in Beijing, China.

Eight (八 *hachi, yaʔ*) is also considered a lucky number in Japanese culture, but the reason is different from Chinese culture. Eight gives an idea of growing prosperous, because the letter (八) broadens gradually.

The Japanese thought eight や *yaʔ* as a holy number in the ancient times and it is thought that it is related to the fact they used eight to express large numbers vaguely like manyfold, やえはたえ *Yae Hatae*ʔ, which is eightfold and twentyfold, many clouds やくも *Yakumo*ʔ, eight clouds, millions and millions of Gods やおよろずのかみ *Yaoyorozu no Kami*ʔ, eight millions of Gods, and so on.

The ancient Japanese also gave importance to pairs, so some researchers guess twice as four, よ *yo*ʔ, which is also guessed to be a holy number in those times because it indicates that the world, north, south, east, and west might be considered a very holy number.

In numerology, **8** is the number of building, and in some theories, also the number of destruction. In astrology, in the Middle Ages, **8** was the number of "unmoving" stars in the sky, and symbolized the perfectioning of incoming planetary energy.

Whether thought of as coincidence or not, the number 8 held a powerful collective meaning as Apollo 8, the second crewed mission in the American Apollo space program became the first spacecraft to break free from the gravitational field of the Earth and be captured by and escape from the gravitational field of another celestial body. Those milestones are the most significant and literally the farthest reaching, most powerful accomplishments of modern man.

Apollo 8 Logo

The number **8** is also reputed to resolve dualities and represent expansion, dissolution, dimension of the timeless, good and bad, right and wrong, day and night, the ability to see and relate to eternal dimensions, balance between forces, connects spirit and matter, developing confidence to follow a vision, and breaks down barriers to transformation, reality, and courage.

Because the number **8** symbolizes power, it is shifted ninety degrees to symbolize infinity, ∞ which refers to something *without any limit,* a concept relevant in a number of fields, predominantly mathematics and physics. Having a recognizable history in these disciplines reaching back into the time of ancient Greek civilization, the term in the English language derives from Latin *infinitas*, which is translated as "unboundedness".

In mathematics, infinity is often treated as a number, but it is not the same kind of number as the real numbers. In number systems incorporating infinitesimals, the reciprocal of an infinitesimal is an infinite number which is a number greater than any real number.

The mathematical symbol for infinity is called the lemniscate. The infinity sign was devised in 1655 by mathematician John Wallis, and named lemniscus, from Latin for ribbon by mathematician Bernoulli about forty years later. The lemniscate is patterned after the device known as a mobius strip, named after Mobius, a nineteenth century mathematician.

Mobius Strip

A mobius strip is a strip of paper that is twisted and attached at the ends, forming an "endless" two dimensional surface. The religious aspect of the infinity symbol predates its mathematical origins. Similar symbols have been found in Tibetan rock carvings; and the ouroboros, or infinity snake, is often depicted in this shape. In the tarot, it represents the balance of forces and is often associated with the magician card.

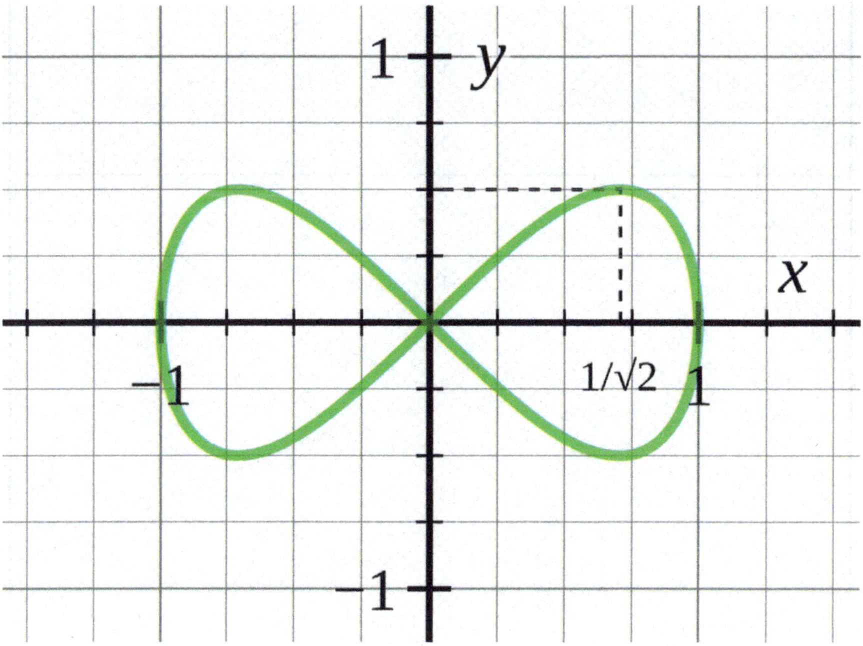

In algebraic geometry, the lemniscate of the figure-eight curve is a plane algebraic curve of degree four and genus zero in the shape of an ∞ symbol defined by the equation: $x^4 - x^2 + y^2 = 0$

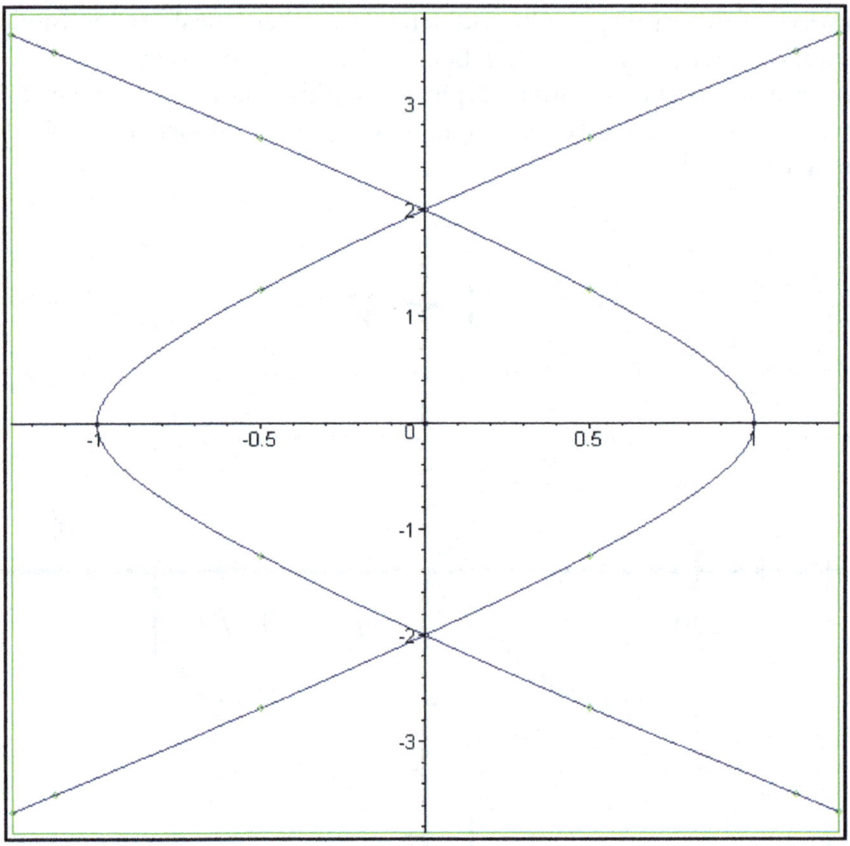

Mathematically the lemniscate is centered on the zero point or void, the transitional center of power consisting of nothing

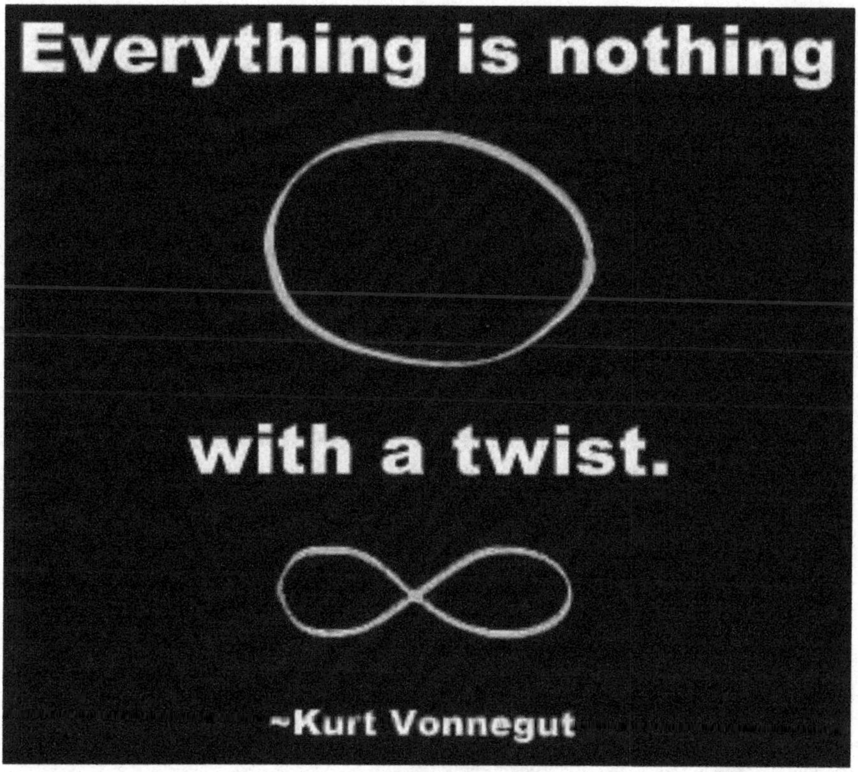
Everything is nothing
with a twist.
~Kurt Vonnegut

THE ZERO POWER POINT OF THE INFINITY ZONE IN THE NATURAL WORLD

The infinity zone is the nexus of power centered at the zero point of the infinity symbol and is prevalent throughout the natural world in perfect form and motion in the grace, beauty, and elegance in the lives of the tiniest insects up through the highest mammals.

Most insects possess wings during part of their life cycles. Insect wings are large folds in the exoskeleton composed of two sheets of cuticle permeated with stiff supportive veins powered by two sets of muscles that independently drive the upstroke and downstroke of the wing movement. The frequency of wing beats ranges from 4 beats per second in butterflies to nearly 1000 beats per second in gnats.

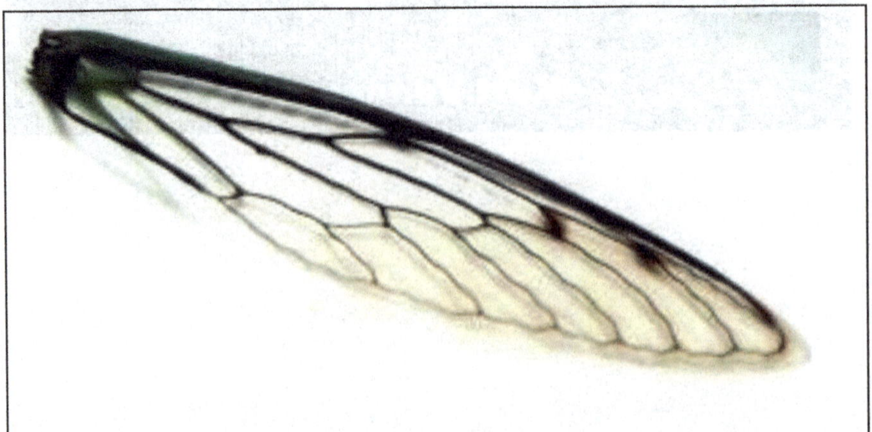

Insect Wing

Insect wings not only move up and down, they move forward and backward in a figure eight pattern.

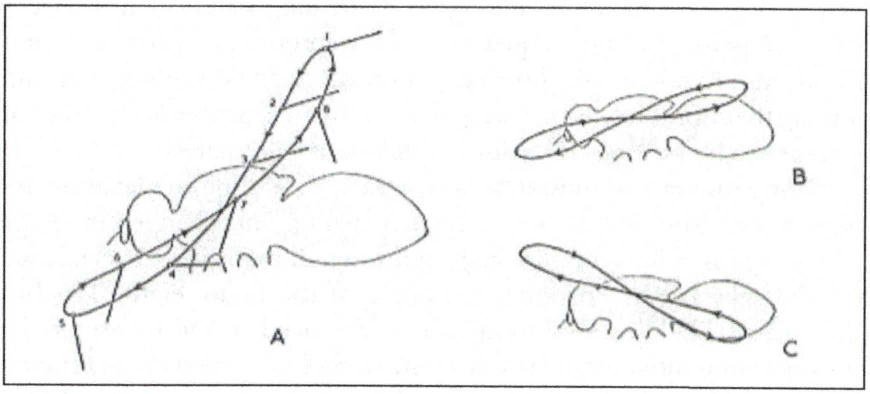

Insect Wing Movements

Given their shape, speed, and stroke pattern, it has never been clearly understood how insect wings generate enough lift to sustain flight. Scientists have discovered that insects generate a vortex, or spiral air motion along the leading edge of their wings. This vortex flows out toward the wing tip in widening spirals and the whirling cylinder of air above the insect provides the extra lift that makes flight possible.

A butterfly has two forewings and two hindwings and when it flexes its thorax muscles to fly it forms a figure eight pattern.

Blue Morpho Butterfly

Insects are the only invertebrates that possess this capability that enables them to exist in great numbers of environmental situations.

The typical insect wing is a superbly designed flying tool capable of a strong sculling action analogous to fanning air into a fire. Insects with two pairs of wings frequently join the anterior and posterior wings by means of hooks and grooves to create a single sculling unit and insects that don't have their wings joined overcome the problem of air turbulence by beating the front and rear pairs alternately.

Wing movement in insects is complex, consisting of elevation and depression, fore and aft movement, twisting, and changes in shape. The wingtips follow a figure eight pattern and many insects can hover or fly backward by changing the angle of the figure eight. The best fliers can fly sideways or rotate about the head or tail by employing unequal wing movement. The wing movement of insects is so efficient that it produces a polarized flow of air from front to rear during 85 per cent of the wingbeat cycle.

The bumblebee is a four-winged **insect with** wings that swivel and flap knocking into each other, producing a buzzing sound. Despite the small area of its wings, the swirling **vortices** produced when the wings collide provide additional lift.

Bumblebee

Bumblebees, like their honeybee cousins, feed on pollen and nectar. A bumblebee often identifies where a food source is by flying in a figure eight pattern around it.

Honeybee workers perform a series of movements, often referred to as the "waggle dance," to teach other workers the location of food sources more than 150 meters from the hive. Scout bees fly from the colony in search of pollen and nectar. If successful in finding good supplies of food the scouts return to the hive and "dance" on the honeycomb.

Honeybee

The honeybee walks straight ahead vigorously shaking its abdomen producing a buzzing sound with the beat of its wings. The distance and speed of this movement communicates the distance of the foraging site to the others. Communicating direction becomes more complex as the dancing bee aligns her body in the direction of the food, relative to the sun. The entire dance pattern is a figure eight with the bee repeating the straight portion of the movement each time it circles to the center again.

In order to hover, a hummingbird's wings move back and forth horizontally, drawing a narrow, elegant figure eight in the air with each full stroke which is continuous like an active Mobius strip moving in a dynamic symbol of infinity. The key to a hummingbird's flight is in its wing patterns. Generally, birds flap their wings up and down. The hummingbird flies by oscillating its wings forward and backward rapidly in a figure eight.

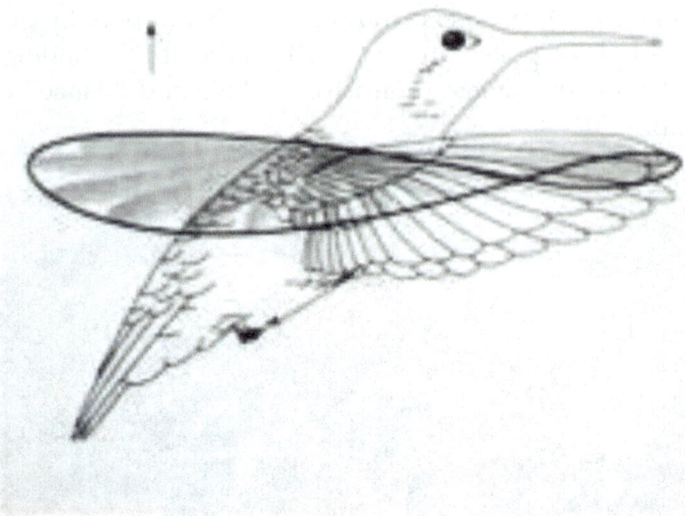

Hummingbird

Seventy five percent of its lift is produced during the down stroke. The figure eight pattern is similar to a swimmer doing figure eights in the water to remain afloat. Making a figure eight with your arms 80 times a second isn't humanly possible, but a hummingbird can. Depending on its size, their wing beat ranges anywhere from 10 to 200 times per second.

A hummingbird's heart is a relatively large organ in comparison to its body weight making up 1.75% to 2.5% of its total weight. The hummingbird's heart is relatively the largest heart in the animal kingdom. It beats about 250 beats per minute at rest and about 1,260 beats per minute while flying.

Dolphins and fish travel through the water by moving their tails in a figure 8 motion, creating figure 8 vortices. It's the fastest, most efficient means of mobility in water.

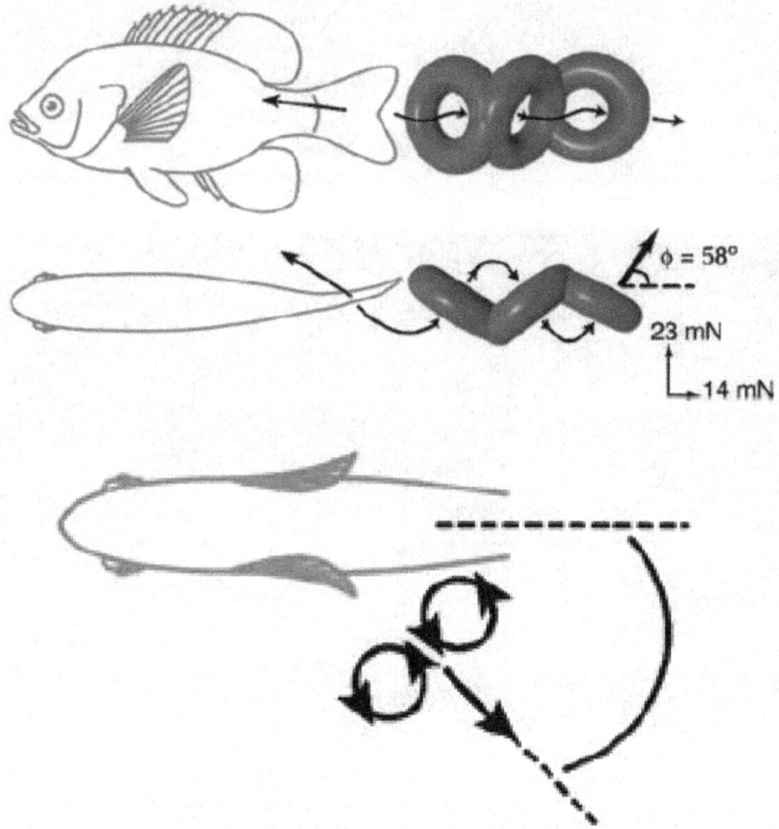

Movement of Fish Tails

When a cheetah sprints, it leaps with both forelegs forward, propelling itself until its front legs meet the back legs, at which point the back legs pick up on the sprint, blasting forward like a second stage booster, propelling its forelegs forward again in a fluid and dynamic figure eight, giving it maximum speed and peak performance.

The gallop of a horse, dogs, wolves, other canines, and the bounding of cheetahs and other cats all move in efficient figure eight patterns that are in harmony with natural forces, giving greater, more precise control. Its natural beauty is self-apparent.

THE INFINITE OCTAVE IN ART, MUSIC, AND THE HUMAN BODY

Art and music reflect the inner life of man, appealing to our innate sense of beauty, inspired by the natural world around us. The geometries present in these disciplines are an integral part of what we perceive as beauty in their expression.

Like the wave to particle transition, when we flow from four dimensional movement back to three dimensional form we dance between inner and outer infinities in a dynamic rapid-fire exchange of subject and object. We recognize the grace of these relationships in nature and mimic its perfection visually in the static arts of painting and photography, and in the dynamic arts of film, dance, sports, and other forms of movement as well as the auditory bliss of divinely inspired music.

In ancient Greece, mathematical disciplines introduced in childhood combined gymnastics and music until age 17 or 18, emphasizing gymnastics until the age of 20. This training was designed to stabilize the equilibrium of the soul, viewed by Plato as straddling the upper Intelligible realm and the lower Sensible realm so the individual could recognize the symmetry and harmony common to both music and bodily movement.

In Egyptian belly dancing a figure 8 with hips is comprised of two different isolations, the hip slide and the twist. The belly dance reverse figure 8 move is called the Turkish figure 8 which requires the dancers to pull in towards the center.

Contra dance choreography specifies the dance formation known as the *figures* and the sequence of those figures in a dance. The figures usually repeat a consistent pattern aligned with the phrasing of the music that typically takes eight counts, although figures with four or sixteen counts are common.

Typical contra dance choreography comprises four parts, either 16 counts or 8 measures long. Most contra dance tunes have two parts, each 8 measures long, each fitting one part of the dance. A contra dance is typically 64 counts with a 32 measure tune.

Choreographed movements in time and space like those associated with dance and music multiply and enhance the activity of the infinity zone of the left and right brain informational volleys and serves across the "net" of the corpus callosum; a flat bundle of neural fibers that connects the left and right hemispheres of the brain.

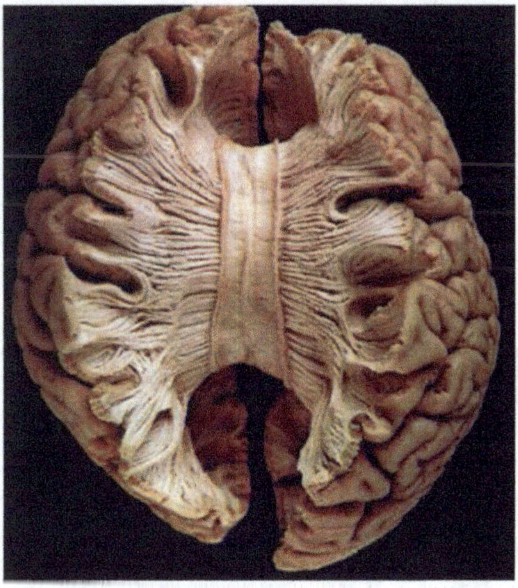

Corpus Callosum

This process is highly active in drumming as described in *The Healing Power of the Drum* by Robert Lawrence Friedman.

"One of the most powerful aspects of drumming and the reason that people have done it since the beginning of being human is that it changes people's consciousness. Through rhythmic repetition of ritual sounds, the body, brain, and the nervous system are energized and transformed. When a group of people play a rhythm for an extended period of time, their brain waves become entrained to the rhythm and they have a shared brain wave state. The longer the drumming goes on, the more powerful the entrainment becomes and it's considered to be the most ancient form of holy communion. All of the oldest known religious rites used drumming as part of the shared religious experience."

Normal outward focused attention generates beta waves that vibrate from 14 to 40 cycles per second. When awareness shifts to an internal focus our brain slows down into the more rhythmical waves of alpha, vibrating at 7-14 waves per second. Alpha is defined by relaxation and centering. Dropping down to 4-7 cycles per second the brain enters the theta state where there is an interfacing of conscious and unconscious processes, producing hypnologic dream-like imagery at the threshold of sleep.

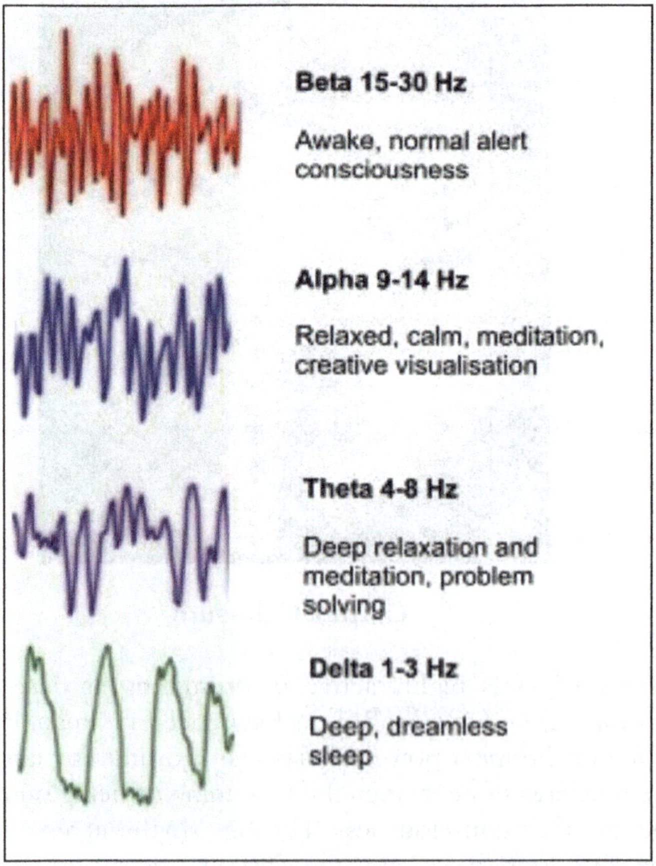

Brain Waves

Theta is the state where mystical insights and creative solutions to complex situations arise and is marked by physical and emotional healing. People with a preponderance of theta brainwaves can learn and process more information than normal. Without some form of intensive training it is hard to stay awake in theta. At this threshold it is easy to slip down into delta, the slowest brainwave frequency of 1-3 cycles per second which is the state of unconsciousness or deep sleep.

The brain is divided into two hemispheres that are split in their control of the thinking process. The right brain functions as the creative, visual, aural and emotional center and the left brain is the rational, logical, analytical, verbal administrator. Generally, either the right or left brain dominates in cycles lasting from 30 minutes to 3 hours.

While one hemisphere is dominant, the memories, skills, and information of the other hemisphere are less available, residing in a subconscious or unconscious realm. Not only do the right and left brain operate in different modes, they usually operate in different brain wave rhythms. The right brain may be generating alpha waves while the left brain is in the beta state, or both can generate the same type of brain waves, but remain out of sync with each other. In states of intense creativity, deep meditation or under the influence of rhythmic sound, both hemispheres can become entrained to the same rhythm. This state of unified whole brain functioning is called hemispheric synchronization or the awakened mind.

Music appeals to our sense of beauty through sound that expresses infinity in time. We appreciate the beauty of music when the musician has mastered the use of octaves, sound, timing, silences, and the universal mathematical laws inherent in its structure. Its appeal is universal.

Pythagoras is often credited for discovering that an oscillating string stopped halfway along its length produces an octave relative to the string's fundamental, while a ratio of 2:3 produces a perfect fifth and 3:4 produces a perfect fourth, but the Chinese had instruments that were thousands of years older like the Guqin, that also feature these tonal scales.

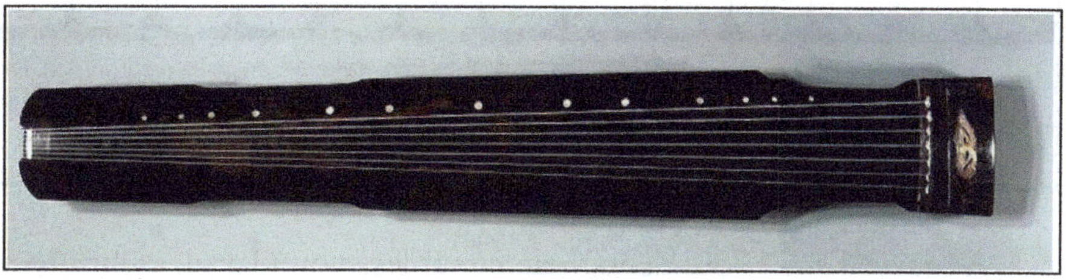

Gugin

Pythagoreans believed that these harmonic ratios gave music powers of healing which could "harmonize" an out-of-balance body, a belief that has been revived in modern times.

In Music:

- A note played for one-eighth the duration of a whole note is called an eighth note, or quaver.
- An octave, the interval between two notes with the same letter name, where one has double the frequency of the other, is so called because there are eight notes between the two on a standard major or minor diatonic scale, including the notes themselves, without chromatic deviation. The ecclesiastical modes are ascending diatonic musical scales of eight notes or tones comprising an octave.
- There are eight notes in the octatonic scale.
- There are eight musicians in a double quartet or an octet. Both terms can also refer to a musical composition for eight voices or instruments.

An octave is the interval in music between two notes where the higher note has twice the frequency of the lower. On a piano keyboard this corresponds to an interval of eight white notes designated C D E F G A B C.

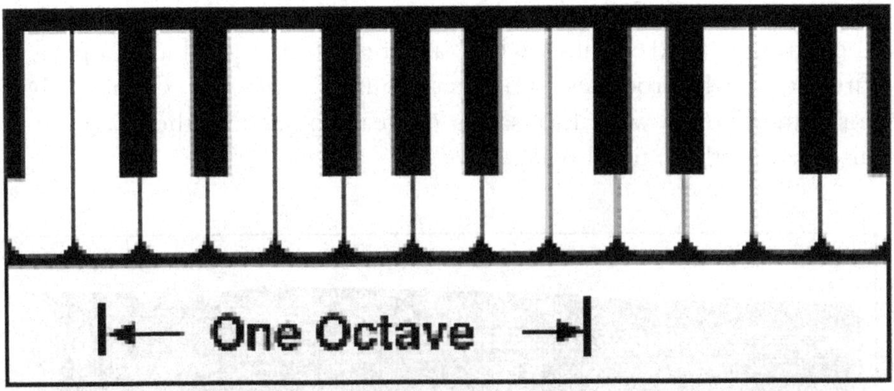

A musical scale sounds incomplete unless you play all eight notes which is why there are eight notes instead of seven in an octave. The term *octave* belongs to Western music, but the musical interval occurs in music around the world.

Harmony is the use of simultaneous pitches, tones, notes, or chords. The study of harmony involves chords, their construction, chord progressions, and the principles of connection that govern them. Harmony is said to refer to the "vertical" aspect of music as distinguished from melodic line, or "horizontal" aspect. Counterpoint refers to the interweaving of melodic lines, and polyphony refers to the relationship of separate independent voices which are sometimes distinguished from harmony.

In popular and jazz harmony chords are named by their root plus various terms and characters indicating their qualities. In many types of music, notably baroque, romantic, modern, and jazz, chords are augmented with "tensions".

A tension is an additional chord member that creates a dissonant interval in relation to the bass. Typically, in the classical Common practice period a dissonant chord, or a chord with tension will resolve to a consonant chord. Harmonization sounds pleasant to the ear when there is a balance between the consonant and dissonant sounds and it occurs when there is a balance between tense and relaxed moments. This concept is known as relaxed tension, which is the central zero point where energy is mastered in balance.

Musica universalis, known as universal music, or music of the spheres, is an ancient philosophical concept that regards proportions in the movements of celestial bodies with the Sun, Moon, and planets as a form of musica, the Medieval Latin name for music. This "music" is not usually thought to be literally audible, but a harmonic and/or mathematical and/or religious concept. The idea appealed to musical thinkers until the end of the Renaissance.

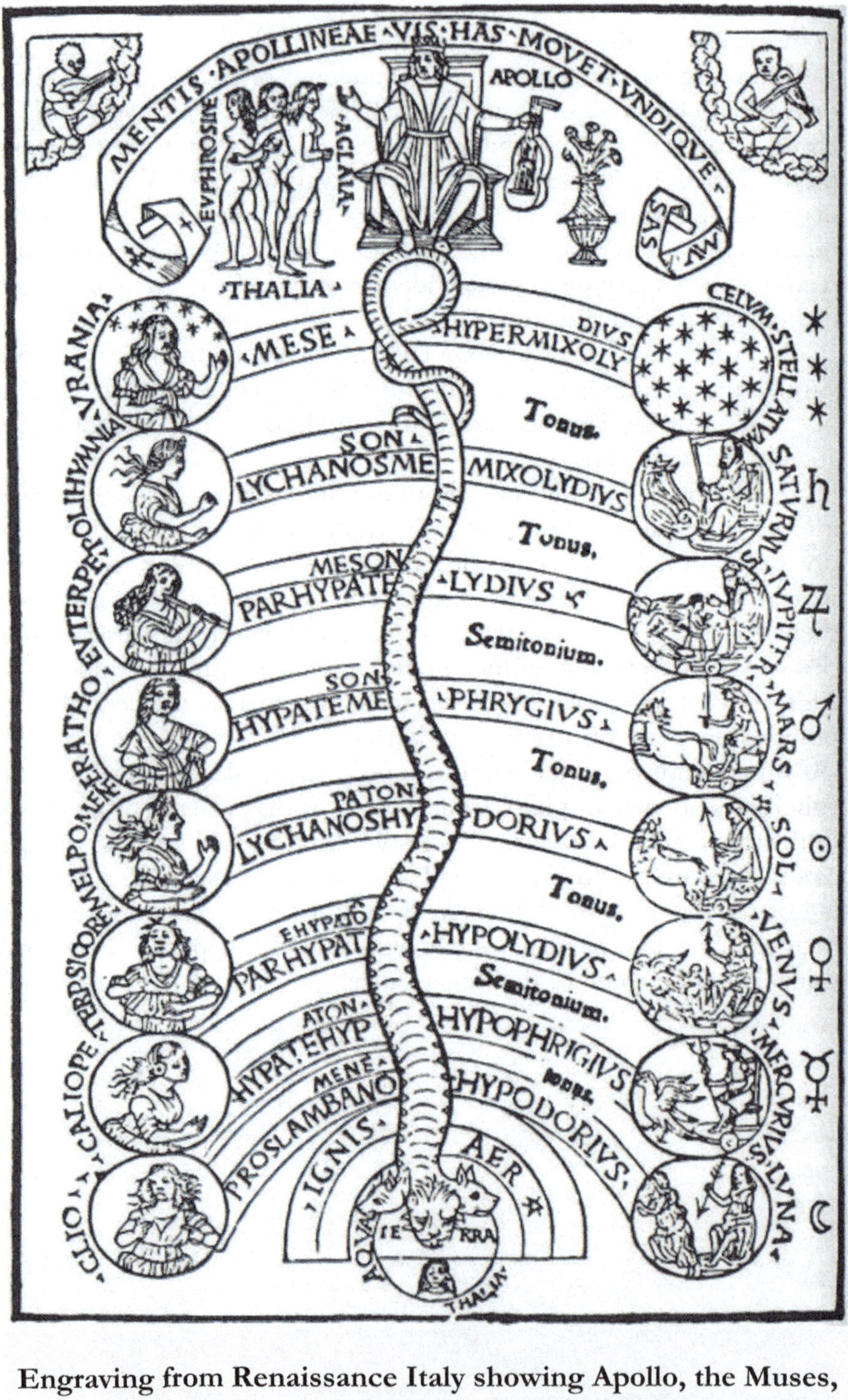

Engraving from Renaissance Italy showing Apollo, the Muses, the planetary spheres, and musical ratios.

The Music of the Spheres incorporates the metaphysical principle that mathematical relationships express qualities or "tones" of energy that manifest in numbers, visual angles, shapes and sounds, all connected within a pattern of proportion.

In a theory known as the Harmony of the Spheres, Pythagoras proposed that the Sun, Moon, and planets all emit their own unique hum known as orbital resonance based on their orbital revolution, and that the quality of life on Earth reflects the tenor of celestial sounds which are physically imperceptible to the human ear. Subsequently, Plato described astronomy and music as "twinned" studies of sensual recognition: astronomy for the eyes and music for the ears, both requiring knowledge of numerical proportions.

Later philosophers kept the close association between astronomy, optics, music, and astrology, including Ptolemy, who wrote influential texts on these topics. Alkindi, in the 9th century, developed Ptolemy's ideas that explored many points of relevance to astrology and the use of planetary aspects.

In the 17th century, Johannes Kepler influenced by arguments in Ptolemy's *Tetrabiblos, Optics and Harmonica,* compiled his *Harmonices Mundi*, "Harmony of the World", that presented his own analysis of optical perceptions, geometrical shapes, musical consonances, and planetary harmonies. According to Kepler, the connection between sacred geometry, cosmology, astrology, harmonics, and music is through *musica universalis*. Kepler regarded this text as the most important work of his career and the fifth part concerning the role of planetary harmony in creation the crown of it. His premise was that as an integral part of Universal Law, mathematical harmony is the key that binds all parts together. One theoretical proposition from his work introduced the minor planetary aspects and harmonics into astrology; another introduced Kepler's third law of planetary motion into astronomy.

The three branches of the Medieval concept of *musica* were presented by Boethius in his book *De Musica*:

- *musica mundana*; sometimes referred to as *musica universalis*
- *musica humana*; the internal music of the human body
- *musica quae in quibusdam constituta est instrumentis;* sounds made by singers and instrumentalists

According to Max Heindel**'s** Rosicrucian writings, the heavenly "music of the spheres" is heard in the *Region of Concrete Thought*, the lower region of the mental plane which is an ocean of harmony. It is also referred to in Esoteric Christianity as the place where the state of consciousness known as the "Second Heaven" occurs.

When this universal order is embraced, the infinite realms that we see outside of ourselves are reflected in the infinite realms inside of ourselves, and the polar aspects of the infinity zone come into resonance, making us one in universal harmony.

This holographic paradigm is what it means to be empowered.

The zero Source power of infinity comes into play in the peak performance of athletes who rely on bodily movement in the same way that dancers flow from three dimensional form to four dimensional movement. The key to peak performance lies in the infinity zone where the groove of relaxed tension brings maximum effectiveness.

In the art of drumming, this same action results in the most effective rim shot on a snare drum when the swing of the arm and the wrist flex upward in a bend, making the first half of a figure 8, then come down, snapping the wrist in a whip like action in the second half, catching the rim and the head of the drum simultaneously, resulting in a powerful "crack" sound.

Another example of this principle occurs in the martial arts where there are many flowing figure 8 movements. One of the most powerful is the basic martial arts punch which starts in a horse stance where the feet are placed a bit wider than the shoulders with the knees bent, lowering the body for greater grounding and balance. With fists clenched at the hips, palms up, the punch begins from the ground up, turning the hips, putting the whole body into it. While rotating the hips in the first movement of a figure 8 the fist comes forward, rotating as it extends the figure 8 of the arm that piggybacks on the energy of the turning hips, adding to its power.

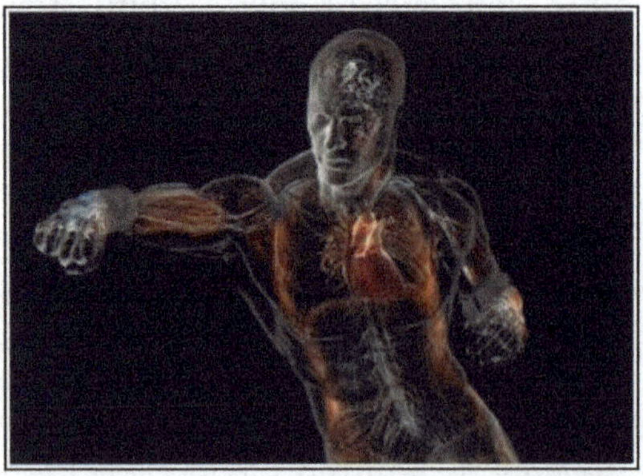

At the end of the punch the hips are turned, with the arm extended at the end of its 180 degree twist. All of the energy is focused on the first two knuckles of the fist, effectively snapping the body and arm like a bullwhip, with the snap concentrated at the tip, completing the first half of the figure 8. The second half finishes on the return to the starting point. In practice both fists alternate in dual figure 8's.

The power center of the human body's infinity zone where the hips turn revolves around what is called the dantian which occurs at the point on the body where the golden ratio makes it cut where the distance between the navel and the foot is taken as 1 unit, and the height of the body is equivalent to 1.618.

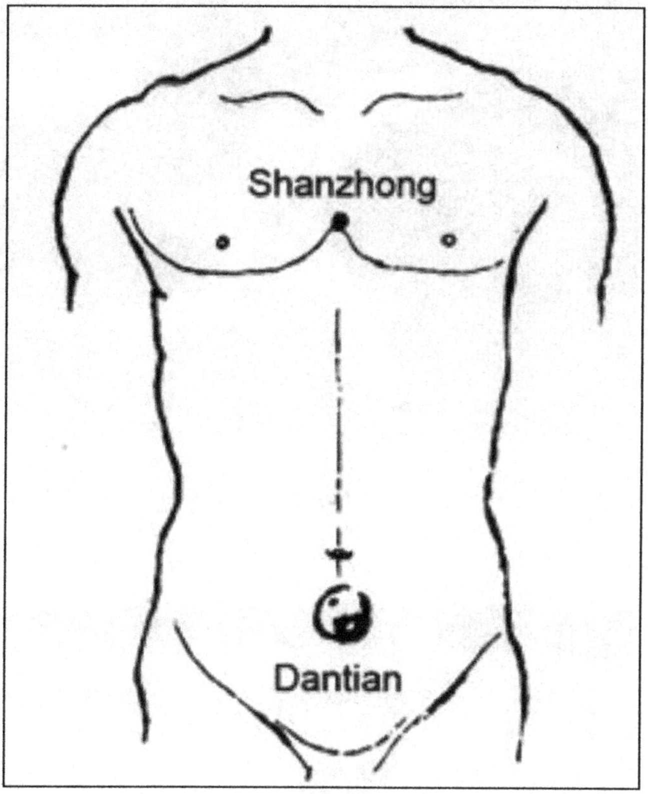

Traditionally, the dantian is considered to be a center of qi or life force energy and is an important point of reference in neidan, qigong, neigong, tao yin, and other self-cultivation practices of exercise, breathing, and meditation as well as in traditional Chinese medicine. The lower dantian is particularly important as the focal point of breathing technique as well as the center of balance and gravity.

Taoist and Buddhist teachers often instruct their students to center the mind in the navel or lower dantian which is believed to aid control of thoughts and emotions and acting from the dantian is thought to be related to higher states of awareness or samadhi.

The Taoist concept of dantians as energy centers is similar to the Indian **yoga** concept of chakras as key points where prana is stored.

When a baseball pitcher throws the ball, the same dynamic is at work with his body and his pitching arm. The movement that comes into play from start to finish in a pitch is a figure 8 that delivers the ball with maximum speed and control.

Ice skaters move their body and legs in alternating figure 8's for optimum velocity and precise control and speed skaters are the masters of this.

Olympic downhill skiers utilize side to side full body figure 8 movements to navigate steep declines with top precision, form, and speed. Those who master it are the ones who win competitions.

Fly fisherman cast their lines in a visible repeating and expanding figure 8 motion that sends the fly sailing to its furthest reaches with the greatest accuracy.

Rowing boats, paddling canoes, and kayaks all use a figure 8 motion similar to the movement of a hummingbird's wings bringing the greatest power into each stroke of the oar or paddle.

The most complex utilization of this infinity zone comes into play in the game of tennis where the movement of the two players brings in multiple interrelated movements along with a dynamic instantaneous flow of projective geometry on physical, mental, and emotional levels.

Rudolf Steiner uncovered geometric principles in his studies of projective geometry and spatial dynamics that laid the groundwork for the figure 8 that comes into play in the flight of the tennis ball and the movement of the racket. Top players raise their hitting arm up as the ball rises over the net, rounding the corner and continuing up into the follow through in a sideways figure 8 manifested in the hitting arm, initiated by a small figure 8 in the hips.

Additionally, a 45 degree angle and its relationship to infinity is everywhere in tennis, beginning with the fact that it is the optimal contact point of every shot in the game. Moving the hips in a figure 8 and keeping a coil in the arms results in a seamless swing. Making contact with the ball this way also alleviates stress or strain in the body, because the hit is absorbed equally throughout the body. Top players understand and feel this synergy of a full-body swing rippling out to the racket head at the point of contact.

This non-linear motion of the infinity zone can be seen throwing or hitting a baseball, throwing or kicking a football, hitting a tennis forehand, backhand, or serve, serving or spiking a volleyball, or driving a golf ball.

This same motion can enhance any athlete's performance.

INFINITY BEYOND THE OCTAVE

The number nine is three squared, the final single digit in our decimal system that precedes the next highest order, ten, symbolized by a one and a zero, and it is the last digit preceding each progression of ten beyond that. Our physical bodies are bilateral, but we have ten toes and ten fingers with an opposable thumb; a physical advantage that ranks us at the top of the earth's animal hierarchy with the higher primates. Physiologically our digits are a prime external indicator of our place at the top of the evolutionary scale, while internally the development of our brain charts the path of the evolution of our consciousness into self-awareness; the capability that distinguishes us above all other life forms on the earth.

The term consciousness refers to common capabilities of humans and other animals, and differentiates between them, designating uniquely human linguistic, rational, and abstract capabilities. Consciousness also includes a range of functions that esoteric traditions claim supersede rational and egoic forms of consciousness representing the evolution of what is conceptualized as spirit, soul, mind, self, and transcendental human capabilities.

Shamanism, the oldest known form of spirituality in the world embraces all of these as well as all other forms of religious thought that followed. A basis for different forms of consciousness underlying shamanic phenomena is found in the major architectural and functional strata of the brain and their different information processes.

Although human consciousness is not specific to any particular function or system of the brain, different modalities of consciousness are associated with different systemic information-processing functions, integration of brain processes, and patterns of homeostasis. All of the major systems of the brain participate in complex human behavior, but specific systemic patterns of brain functioning are associated with distinct experiential states and modes of consciousness.

Recognition that consciousness is tied to the functioning of a biological system does not require reduction of consciousness solely to the functions of the biological system. A neurophenomenological approach illustrates that both epistemic constructs and physiological patterns of brain operation contribute to consciousness. The relationship of brain physiology to consciousness is illustrated through an examination of how the physical structures of the brain and their associated activities relate to patterns of consciousness.

Shamanic altered states of consciousness evoke communicative responses from the paleomammalian brain that provide the basis for an expansion of consciousness by integrating information from the lower brain systems into operational activities of the frontal brain establishing synchrony with the frontal brain that permits symbolic reprogramming of the emotional dynamics and behavioral repertoires of lower brain centers.

A fully turned on 3 X 3 brain could produce a shift from our present fourth dimensional existence that perceives a three dimensional world to a possible fifth dimension where we perceive the fourth dimension, transcending time and duality. A transcendence from 2^2, 2 X 2, the model for three dimensional existence to the next logical progression of 3^2, 3 X 3, the number 9 could provide a new model for the fifth dimension which has been referred to as the second coming of Christ consciousness, the return of Quetzalcoatl, and other mythologies from diverse cultures.

Aside from continuing breakthroughs in quantum physics that prove that material reality is in fact consciousness, there can be no doubt that profound changes are upon us. Global warming is a fact, as is a shift in the magnetic poles of the earth, and an upward shift in the internal frequency of our planet.

The earth's frequency was previously thought to be constant. When NASA started measuring it, it was 7.8 hertz, but it has been measured as high as 14, and it is speeding up. At the same time the field strength of the earth's gravity is dropping at an alarming rate. In the last few years magnetic north has been moving rapidly, and the inner core of the earth has displaced itself from the angle of rotation of the earth's crust. The sun's own magnetic north and south is no longer detectable as of 1995 and solar flares and proton storms are off the scale.

On December 22, 2012 the sun was in alignment with the galactic equator of the Milky Way Galaxy which occurs every 25,800 years and was predicted by the Mayans. This is also the date of the birth of Venus which represents the feminine, the right brain intuitive energy that has been suppressed for far too long.

Energetically all of these phenomena represent an electromagnetic shift upward. The results of what occurs from an increase in frequency were proven in Cymatics, the study of wave phenomena pioneered by Swiss medical doctor and natural scientist, Hans Jenny who conducted experiments animating inert powders, pastes, and liquids into life-like flowing forms that mirrored patterns found throughout nature, art, and architecture. These patterns were created with sine wave vibrations using pure tones within the audible range so observers could see a physical representation of vibration, or how sound manifests into form through the medium of different materials.

Cymatics shows that when vibrational patterns were produced in series and compared, the same formal pattern recurred at increasing frequencies. The number of constituent elements also increased at higher frequencies. At a particular frequency a specific matrix pattern forms, and with an increase in frequency the pattern dissolves into chaos before reforming into a more complex pattern at a higher frequency.

In other words chaos precedes reformation from a lower order into a more complex higher order. The old structure has to break down in order for the new more complex one to form. As humans, when we are born into this world we go from amniotic bliss through the chaos and violence of the birth process until we are born anew into a whole new paradigm of being.

Numerous unprecedented changes are occurring within and around us and we are feeling our subjective experience of time moving faster and faster because our consciousness is quickening in preparation for what could be an upcoming dimensional shift. If we are aware enough to be in harmony with the natural unfolding cosmic laws of oneness, reciprocity, and balance, and we are in harmony and alignment within ourselves, living in full integrity and unity with all that is, we may ride the wave and flow forward with the ongoing shift to the next higher dimension along with our mother earth, and we just might become fully conscious co-creators of our holographic reality.

PORTAL TO THE INFINITE

Infinity is central to the great cosmic mystery that pushes the limits of everything we know, which by its very nature is finite. When we try to define it, infinity loses its meaning by the very attempt at definition, which tries to contain what cannot be bounded. It is a paradox seen in our "inner" infinity zone where our consciousness is centered in a place that contains power, mystery, and universal wisdom that becomes apparent through the perception of higher consciousness.

Unseen yet perceived, this implicate interference pattern defines the ancient wisdom of, "as above, so below", and though unseen and pointing toward the unknown, it resides within us; the living, breathing, dynamic manifestation of the nexus between the third and fourth dimensions of time and space.

The center of the Universe is right between our eyes as well as everywhere else outside of us.

Though physically intangible, the magic of its power is available to us by understanding and cultivating its combined properties of dynamic form and motion through our body, the vehicle of consciousness, the vessel that carries us on our journey through the third dimension.

Like the point with no dimension it is the place where spirit meets matter.

In the world as we know it space is the final frontier and the mind's capacity to observe and experience a more holistic conception of it is the quest. Rudolf Steiner described a state of consciousness where a different kind of space is experienced that is the polar opposite to our ordinary Euclidian conception. This consciousness looks in from the periphery toward an infinite inwardness in contrast to the "normal" experience that looks out from a center toward an infinite outwardness following a universal holographic order centered on zero that comes from nothing, giving reality to the paradox that everything is nothing and nothing is everything.

Expanding and contracting spirals and geometric progressions have their basis in the divine proportion, the mathematical representation of a hologram that manifests in endless permutations through every day visible expressions and microcosmic quantum realms and the deeper we delve into these puzzling realities the more bizarre and unpredictable they become. We live at the center of a Great Mystery and in spite of our advanced technologies we are no closer to the truth. Our sophisticated technologies driven by our divide and conquer mentality have only succeeded in isolating us from the natural world even more.

Moving from esoteric concerns into the realm of the biological, we fear what we can't comprehend and act like little children who are afraid of the dark until we become en*light*ened, proving and reinforcing the fact that fear is contraction and love is expansion.

Faced with the Great Mystery of our existence we can "bury our heads in the sand" and continue losing ourselves and the world we live in through the mirrors of our narcissistic cell phone and internet consumer culture, drugs, alcohol, and any number of other addictions and distractions that we have created with the transitory "magic" of our technology, reinforcing our distance from the natural world, or we can face the grand mystery of something much bigger than we are.

If we look deep within ourselves and see the above and below of our inner and outer worlds and realize that we have the unique position of straddling them both we can realize the perfect reflection of holographic truth that we exist within.

Regardless of the path we take, if we follow the unfolding self-evident maps, where might they take us?

What happens when our short lived sojourn here on the planet earth ends?

We can only speculate what death brings in our release from three dimensional reality and we can push the limits of our consciousness through exploratory altered states the way shamans have since time immemorial, cracking open inner space and storming heaven to the best of our abilities, but regardless of what we may learn and experience, what happens after we die is still a mystery that will only be revealed when we cross that threshold - alone.

Regardless of the paradoxical mystery of the Source, we owe it to ourselves to consider any possibilities that present themselves and if they have the ring of truth there is no harm in acknowledging the holographic reality that resonates with us at the core of our existence.

Bibliography

Abrams, David, *The Spell Of The Sensuous: Perception And Language In a More-Than -Human World*, (New York: Random House, 1996)

Campbell, Joseph, *The Hero With A Thousand Faces*, (Novato: New World Library)

Doreal, Dr. M., *The Emerald Tablets of Thoth The Atlantean.* (Nashville: Source Books, 2002)

Eliade, Mircea, *Shamanism, Archaic Techniques of Ecstasy*, (Princeton: University Press, 2004)

Friedman, Robert L., *The Healing Power of the Drum,* (Philadelphia: White Cliffs Media Co, 2000)

Grof, Stanislaus, & Bennet, Hal, *The Holotropic Mind: The Three Levels of Human Consciousness and How They Shape Our Lives*, (New York: Harper Collins.1993)

Haule, John R., *Taking Direction from the Spirit in Shamanism and Psychotherapy*, (Shamanic Applications Review 1997)

Havelock, Eric A., *The Muse Learns to Write,* (New Haven: Yale University Press, 1988)

Jenny, Hans, *Cymatics: A Study of Wave Phenomena and Vibration*, (Newmarket: MACROmedia, 2001)

Jung, Carl, *The Portable Jung,* (New York: Penguin, 1976)

Jung, Carl, *The Psychology of the Transference,* (Princeton: Princeton University Press, 1959)

Jung, Carl, *The Undiscovered Self,* (New York: Signet Books, 2006)

Jung, Carl, *Symbols of Transformation*, (New York: Pantheon, 1956)

Kenyon, Tom, *The Hathor Material: Messages From an Ascended Civilization*, (Baltimore: Orb Communications, 2012)

Lawlor, Robert, *Sacred Geometry: Philosophy & Practice,* (London: Thames & Hudson, 1982)

MacLean, P., *The Triune Concept of Brain and Behavior,* (Toronto: University of Toronto Press, 1973)

MacLean, P., *The Triune Brain in Evolution,* (New York: Plenum Press, 1990)

McKee, Robert, *Story,* (New York: Regan Books, 1997)

McKenna, Terence, *Food of the Gods,* (New York: Bantam Books, 1992)

McNeley, James K., *Holy Wind in Navajo Philosophy*, (Tucson: University of Arizona Press, 1981)

Merkur, Dan, *Becoming Half Hidden: Shamanism and Initiation Among the Inuit*, (New York: Garland Publishing, 1992)

Merleau-Ponty, Maurice, *Phenomenology of Perception*, (London: Routledge & Kegan Paul, 1962)

Monbiot, George, Regenesis: *Feeding the World Without Devouring the Planet*, (New York: Penguin Random House, 2022)

Nicoll, Maurice. *Psychological Commentaries on the Teachings of Gurdjieff & Ouspensky*. London: Watkins, 1980.

Ouspensky, Pyotr D., *In Search of the Miraculous*. (San Diego: Harcourt Brace, 1949)

Pallamary, Matthew, J., *Spirit Matters*, (San Diego: Mystic Ink Publishing, 2007)

Pallamary, Matthew, J., *The Center Of The Universe is Right Between Your Eyes, But Home Is Where The Heart Is*, (San Diego: Mystic Ink Publishing, 2017)

Pallamary, Matthew, J., *PICAFLOR*, (San Diego: Mystic Ink Publishing, 2021)

Pallamary, Matthew, J., *PICAFLORES: The Nerve Endings of GOD*, (San Diego: Mystic Ink Publishing, 2022)

Pallamary, Matthew, J., Mayberry, Paul, *The Infinity Zone*, (San Diego: Mystic Ink Publishing, 2012)

Schwaller de Lubicz, R. A., *The Temple in Man*, (Rochester: Inner Traditions, 1981)

Schwaller de Lubicz, R. A., *The Temple of Man*, (Rochester: Inner Traditions, 1998)

Sheldrake, Rupert, Morphic Resonance: The Nature of Formative Causation, (Rochester: Park Street Press, 2009)

Steiner, Rudolph, *The Fourth Dimension: Sacred Geometry, Alchemy, and Mathematics*, (Great Barrington: Anthroposophic Press, 2001)

Stevens, José, *Earth to Tao*, (Santa Fe: Bear and Company, 1994)

Talbot, Michael, *The Holographic Universe*, (New York: Harper Perennial, 2011)

Thomas, Nick. *Science Between Space And Counterspace*. London: New Science Books, 1999.

Winkelman, Michael, *Shamanism: The Neural Ecology of Consciousness and Healing*, (Westport: Bergin & Garvey, 2000)

URLS

https://mattpallamary.com
https://mysticinkpublishing.com

Counter Space
http://www.nct.anth.org.uk/counter.htm
Dance
http://en.wikipedia.org/wiki/Contra_dance_choreography

Fish
http://www.people.fas.harvard.edu/~glauder/reprints_unzipped/FishLauder2006.pdf

Harmony
http://en.wikipedia.org/wiki/Harmony

Hummingbirds
http://www.ehow.com/how-does_4566089_a-hummingbird-fly.html#ixzz1nRkn8szJ
http://www.birdfeeders.com/resources/wild-bird-articles/hummingbirds/to-catch-a-hummingbird
http://www.nectarartprints.com/hb_flight.htm
http://www.hummingbirds.net/images/figure8.mov
http://clackhi.nclack.k12.or.us/physics/projects/rates_around_us/hummingbird%20Folder/hummingbirds.htm
http://hummingbirdworld.com/h/behavior.htm
http://www.youtube.com/watch?v=x3th_3BZVHQ
http://www.youtube.com/watch?v=ils4cqeekSI
http://www.bbc.co.uk/nature/life/Hummingbird#p00cn9h7
http://www.instructables.com/id/Figure-Eight-Motion-Drive/
http://jeb.biologists.org/content/210/13/2368.full.pdf
http://blogs.bu.edu/biolocomotion/2011/10/21/its-a-bird-no-its-an-insect-no-wait-its-a-hummingbird/

Infinity
http://en.wikipedia.org/wiki/Infinity
http://www.goodreads.com/quotes/tag/infinity
http://f06.middlebury.edu/FYSE1176A/Infinity_Quotes.html

Insects

http://www.everythingabout.net/articles/biology/animals/arthropods/insects/
http://www.insectman.us/articles/misc-design/flight/testimony-creation.htm
http://www.bumblebee.org/bodyEyehtm.htm
http://insects.about.com/od/antsbeeswasps/p/honeybeecommun.htm
http://www.ehow.com/facts_5180742_butterfly-wings.html

Music
http://en.wikipedia.org/wiki/Musica_universalis

Number 8
http://www.virtuescience.com/8.html
http://en.wikipedia.org/wiki/8_(number)
http://richardphillips.org.uk/number/Num8.htm
http://www.ridingthebeast.com/numbers/nu8.php
http://mysticalnumbers.com/Number_8.html
http://numerology-thenumbersandtheirmeanings.blogspot.com/2011/05/number-8.html
http://en.wikipedia.org/wiki/Apollo_8
Sacred Geometry
http://en.wikipedia.org/wiki/Sacred_geometry
http://www.goldennumber.net/
http://milan.milanovic.org/math/english/body/
http://www.software3d.com/Stella.php

Additional URL Refences

https://aleteia.org/2022/02/07/science-confirms-the-importance-of-religions-tree-of-life/
https://www.ancient-code.com/scientists-on-how-trees-talk-through-an-ancient-otherworld-network/
https://www.ancient-code.com/12-sacred-trees-of-world-mythology-and-their-meaning/
https://bigthink.com/hard-science/the-universe-works-like-a-huge-human-brain-discover-scientists/
https://bigthink.com/starts-with-a-bang/universe-fractal/
https://www.cambridge.org/core/journals/international-journal-of-astrobiology/article/intelligence-as-a-planetary-scale-

process/5077C784D7FAC55F96072F7A7772C5E5
https://www.collective-evolution.com/2020/02/24/study-the-human-body-responds-to-sharp-changes-in-solar-geomagnetic-activity/?fbclid=IwAR292niSxBCLGr9lu995Wa7cjpmxOMKtrG2PGimOxffOamqP-F-m4pHyl3M
https://www.collective-evolution.com/2014/11/24/what-science-is-telling-us-about-the-hearts-intuitive-intelligence/
https://cosmosimfrazza.myfreesites.net/cosmic-web-and-brain-network-datasets
https://cosmometry.net/fractal-holographic-synergetic-universe.html
https://cosmometry.net/overview-of-cosmometry.html
https://directorsblog.nih.gov/2021/08/03/the-amazing-brain-toward-a-wiring-diagram-of-connectivity/?fbclid=IwAR3KRrcjjeb24DvKTbyLkbksD65NjMCvH4grlHkfaSm8U3IHqso5H8HrRU4
https://en.wikipedia.org/wiki/Golden_ratio
https://en.wikipedia.org/wiki/Wave%E2%80%93particle_duality?wprov=sfla1
https://en.wikipedia.org/wiki/Symbiosis
https://www.eurekalert.org/news-releases/834792
https://www.frontiersin.org/articles/10.3389/fphy.2020.525731/full
https://futurism.com/physicist-entire-universe-neural-network
https://futurism.com/the-byte/microsoft-researchers-claim-entire-universe-is-a-machine-learning-algorithm
https://www.gaia.com/article/mysteries-of-the-human-heart
https://www.goldennumber.net/solar-system/#:~:text=The%20dimensions%20of%20the%20Earth,and%20the%20Moon%20to%20scale.
https://www.heartmath.org/articles-of-the-heart/the-math-of-heartmath/heart-intelligence/
https://www.heartmath.org/research/science-of-the-heart/
https://heart-saverinstitute.com/heart-intelligence-the-heart-is-more-powerful-than-the-brain/
https://www.independent.co.uk/life-style/gadgets-and-tech/universe-brain-shape-cosmic-web-galaxies-neurons-b1724170.html
https://www.livescience.com/is-the-universe-conscious.html
https://makeupandbreakup.com/2019/12/12/mathematics-is-the-language-of-universe/
https://www.microsoftnewskids.com/en-us/kids/animals/ants-observed-healing-wounded-trees-in-bizarre-never-before-seen-

behavior/ar-AASwkt6?ocid=msedgdhp&pc=U531
http://m.nautil.us/issue/50/emergence/the-strange-similarity-of-neuron-and-galaxy-networks
https://www.physicsoftheuniverse.com/topics_quantum_nonlocality.html
https://physicsworld.com/a/is-the-universe-a-dodecahedron/
https://phys.org/news/2020-11-human-brain-resemble-universe.html
https://www.popularmechanics.com/science/a36112655/universe-is-self-learning-algorithm/?utm_campaign=socialflowFBPOP&utm_source=facebook&utm_medium=social-media&fbclid=IwAR1mWovziRASDzmRP8aoQV3qwgwpu-2xbmecOFcctrzlIQEtYmq0siyXXCs
https://thepsychologist.bps.org.uk/grand-unified-theories-consciousness
https://riskgroupllc.com/universal-geometric-language/#:~:text=Since%20mathematical%20models%20can%20define,is%20also%20common%20to%20everyone
https://www.robertlanzabiocentrism.com/how-we-collectively-determine-reality-and-the-structure-of-spacetime-itself/
https://www.sacredgeometrycentral.com/golden-ratio
https://sacredgeometryinternational.com/does-the-human-brain-resemble-the-universe/?mc_cid=f193dd77c7&mc_eid=1b1f53093f
https://www.recoveredscience.com/const305goldendrawings.htm
https://www.sciencealert.com/wildly-fun-new-paper-compares-the-human-brain-to-the-structure-of-the-universe
https://www.sciencedirect.com/science/article/pii/S1002007108003419
https://www.solargeometry.com/
https://www.syfy.com/syfy-wire/is-our-brain-a-mirror-of-the-universe
https://www.thebrighterside.news/post/for-the-first-time-ever-scientists-create-matter-directly-from-collisions-of-light
https://theconversation.com/pythagoras-revenge-humans-didnt-invent-mathematics-its-what-the-world-is-made-of-172034
https://www.theguardian.com/environment/2022/may/07/secret-world-beneath-our-feet-mind-blowing-key-to-planets-future
https://www.goldennumber.net/architecture/
https://www.goldennumber.net/triangles/
https://thenextweb.com/news/physicists-working-with-microsoft-

think-the-universe-is-a-self-learning-computer
https://www.universetoday.com/148966/one-of-these-pictures-is-the-brain-the-other-is-the-universe-can-you-tell-which-is-which/
https://unyte.com/blogs/news/how-your-heart-may-be-your-wisest-brain
https://www.vice.com/en/article/bvxm83/human-brain-and-universe-look-similar-says-study
https://www.zukunftsstiftung-entwicklung.de/media/Bilder_ZSE/UEber_Uns_Dateien/Grundlagentexte/PLANT_STUDY_MANUAL.pdf

ABOUT THE AUTHOR

Matthew J. Pallamary's works have been translated into Spanish, Portuguese, Italian, Norwegian, French, and German. His historical novel of first contact between shamans and Jesuits in 18th century South America, titled, ***Land Without Evil*** received rave reviews along with a San Diego Book Award for mainstream fiction. It was also adapted into a full-length stage and sky show, co-written with and directed by Agent Red and performed by Sky Candy, an Austin Texas aerial group. The making of the show was the subject of a PBS series, Arts in Context episode, which garnered an EMMY nomination.

His nonfiction book, ***The Infinity Zone: A Transcendent Approach to Peak Performance*** is a collaboration with professional tennis coach Paul Mayberry that offers a fascinating exploration of the phenomenon that occurs at the nexus of perfect form and motion. ***The Infinity Zone*** took 1st place in the International Book Awards, New Age category and was a finalist in the San Diego Book Awards.

His first book, a short story collection titled ***The Small Dark Room Of The Soul*** was mentioned in The Year's Best Horror and Fantasy and received praise from Ray Bradbury and has been released as an audio book.

His second collection, ***A Short Walk to the Other Side*** was an Award Winning Finalist in the International Book Awards, an Award Winning Finalist in the USA Best Book Awards, and an Award Winning Finalist in the San Diego Book Awards. It has been released as an audio book.

DreamLand a novel about computer generated dreaming, written with legendary DJ Ken Reeth won first place in the Independent e-Book Award in the Horror/Thriller category and was an Award Winning Finalist in the San Diego Book Awards. It has also been released as an audio book.

It's sequel, ***n0thing*** is titled after the main character, who in the real world is his nephew, an international Counter-Strike gaming champion. After winning what amounts to the Super Bowl of gaming, n0thing and his winning teammates, are recruited as a literal "dream team" whose mission is to go into the nightmares of battle scarred veterans and rescue them from their traumatic memories while becoming ambassadors for a gaming platform that exceeds virtual reality with an experience that pushes the boundaries of reality itself.

Eye of the Predator was an Award Winning Finalist in the Visionary Fiction category of the International Book Awards. ***Eye of the Predator*** is a supernatural thriller about a zoologist who discovers that he can go into the minds of animals.

CyberChrist was an Award Winning Finalist in the Thriller/Adventure category of the International Book Awards. ***CyberChrist*** is the story of a prize winning journalist who receives an email from a man who claims to have discovered immortality by turning off the aging gene in a 15 year old boy with an aging disorder. The forwarded email becomes the basis for an online church built around the boy, calling him CyberChrist. It has also been released as an audio book.

Phantastic Fiction – A Shamanic Approach to Story took first place in the International Book Awards Writing/Publishing category. ***Phantastic Fiction*** is Matt's guide to dramatic writing that grew out of his popular Phantastic Fiction Workshop.

Night Whispers was an Award Winning Finalist in the Horror category of the International Book Awards. Set in the Boston neighborhood of Dorchester, ***Night Whispers*** is the story of Nick Powers, who loses consciousness after crashing in a stolen car and comes to hearing whispering voices in his mind. When he sees a homeless man arguing with himself, Nick realizes that the whispers in his head are the other side of the argument.

His memoir ***Spirit Matters*** detailing his journeys to Peru, working with shamanic plant medicines took first place in the San Diego Book Awards Spiritual Book Category, and was an Award-Winning Finalist in the autobiography/memoir category of the National Best Book Awards.

The Center Of The Universe Is Right Between Your Eyes But Home Is Where The Heart Is was an Award Winning Finalist in the International Book Awards. Based on a lifetime of research into shamanism, visionary states, the evolution of written communication and the roots of storytelling, award-winning author, editor, and shamanic explorer Matthew J. Pallamary takes those with open minds courageous enough to question the illusions that most of us think of as real on an expansive journey that pierces the veil of reality itself.

AfterLife: The Adventures of a Lost Soul was inspired by real life events, William Peter Blatty's ***The Exorcist***, and the dynamics of demonic possession.

Matt has also produced and directed ***The Santa Barbara Writers Conference Scrapbook*** documentary film and co-wrote the book of the same title in collaboration with Y. Armando Nieto, and conference founder Mary Conrad.

Death: (A Love Story) a first person narrative spoken by the omniscient voice of Death itself, who says, "I'm here to tell you stories and share some science, history, and myths, all of which are your creations that I want to share to help you understand me more. You have seen me as Satan, Anubis, Mot, Thanatos, God, the Devil, loving, punitive, dark, light – the list goes on and on! It is my sincerest hope that our friendly reintroduction here will change the way you think of me, and maybe in some small way reflect the depth of the love I have for you.

Picaflor is the sequel to ***Spirit Matters***, a San Diego Book Award winner and an Award-Winning Finalist in the National Best Book Awards that chronicles the two decades since of Matthew (Mateo) J. Pallamary's adventures in ***Spirit Matters*** through the mountains, deserts, and jungles of North, Central, and South America pursuing his studies of shamanism and visionary experience working with plant medicines and shamanic plant diets, among them Ayahuasca, Peyote, San Pedro cactus, and many more.

Matt's work has appeared in Oui, New Dimensions, The Iconoclast, Starbright, Infinity, Passport, The Short Story Digest, Redcat, The San Diego Writer's Monthly, Connotations, Phantasm, Essentially You, The Haven Journal, The Hurricanes & Swan Songs Anthology, The Santa Barbara Literary Journal, The Closed Eye Open, The Montecito Journal, and many others. His fiction has been featured in The San Diego Union Tribune which he has also reviewed books for, and his work has been heard on KPBS-FM in San Diego, KUCI FM in Irvine, television Channel Three in Santa Barbara, and The Susan Cameron Block Show in Vancouver. He has been a guest on the following nationally syndicated talk shows; Coast to Coast with George Noory, Paul Rodriguez, In The Light with Michelle Whitedove, Susun Weed, Medicine Woman, Inner Journey with Greg Friedman, and Environmental Directions Radio series. Matt has appeared on the following television shows; Bridging Heaven and Earth, Elyssa's Raw and Wild Food Show, Things That Matter, Literary Gumbo, Indie Authors TV, and ECONEWS. He has also been a frequent guest on numerous podcasts, among them, The Psychedelic Salon, Black Light in the Attic, Third Eye Drops, C-Realm, Psychedelics Today, Voices in the Dark, Adventures Through the Mind, Beyond the Veil, and many others.

Matt received the Man of the Year Award from San Diego Writer's Monthly Magazine and has taught a fiction workshop at the **Southern California Writers' Conference** in San Diego, Palm Springs, and Los Angeles, and at the **Santa Barbara Writers' Conference** for over thirty years. He has lectured at the Greater Los Angeles Writer's Conference, the Getting It Write conference in Oregon, the Saddleback Writers' Conference, the Rio Grande Writers' Seminar, the

National Council of Teachers of English, The San Diego Writer's and Editor's Guild, The San Diego Book Publicists, The Pacific Institute for Professional Writing, The 805 Writers Conference, The College of Central Florida, Yakima Valley College in Washington, The Yakima Public School System, and he has been a panelist at the World Fantasy Convention, Con-Dor, and Coppercon. He is presently Editor in Chief of Mystic Ink Publishing.

Matt was a featured lecturer and performer at the **Mysteries of the Amazon** exhibit at the Appleton Museum in Ocala Florida and The Larson Gallery in Yakima Washington. He frequently visits the mountains, deserts, and jungles of North, Central, and South America pursuing his studies of shamanism.

MATTPALLAMARY.COM

BOOKS BY MATTHEW J. PALLAMARY

THE SMALL DARK ROOM OF THE SOUL

LAND WITHOUT EVIL

SPIRIT MATTERS

DREAMLAND (WITH KEN REETH)

THE INFINITY ZONE (WITH PAUL MAYBERRY)

A SHORT WALK TO THE OTHER SIDE

CYBERCHRIST

EYE OF THE PREDATOR

PHANTASTIC FICTION

NIGHT WHISPERS

THE SANTA BARABARA WRITERS CONFERENCE SCRAPBOOK (WITH MARY CONRAD & Y. ARMANDO NIETO)

n0THING

AFTERLIFE: THE ADVENTURES OF A LOST SOUL

THE CENTER OF THE UNIVERSE IS RIGHT BETWEEN YOUR EYES BUT HOME IS WHERE THE HEART IS

DEATH: (A LOVE STORY)

PICAFLOR

PICAFLORES: THE NERVE ENDINGS OF GOD

Made in United States
Orlando, FL
04 November 2026

90201512R00164